T0321860

Two-Dimensional Information Theory and Coding

This self-contained introduction to two-dimensional (2-D) theory and coding provides the key techniques for modelling data and estimating their information content. Throughout, special emphasis is placed on applications to transmission, storage, compression, and error protection of graphic information.

The book begins with a self-contained introduction to information theory, including concepts of entropy and channel capacity, which requires minimal mathematical background knowledge. It then introduces error-correcting codes, particularly Reed–Solomon codes, the basic methods for error-correction, and codes applicable to data organized in 2-D arrays. Common techniques for data compression, including compression of 2-D data based on application of the basic source coding, are also covered, together with an advanced chapter dedicated to 2-D constrained coding for storage applications.

Numerous worked examples illustrate the theory, whilst end-of-chapter exercises test the reader's understanding, making this an ideal book for graduate students and also for practitioners in the telecommunications and data-storage industries.

JØRN JUSTESEN is a Professor in the Department of Photonics Engineering at the Technical University of Denmark (DTU), a position he has held since 1976. He has previously held visiting positions at the Institute for Information Transmission Problems, Moscow, and the University of Maryland, College Park.

SØREN FORCHHAMMER is an Associate Professor in the Department of Photonics Engineering at DTU. He has previously held visiting positions at IBM Almaden Research Center, California, and at McMaster University, Ontario.

Two-Dimensional Information Theory and Coding

With Application to Graphics and High-Density Storage Media

Jørn Justesen and Søren Forchhammer

Technical University of Denmark

CAMBRIDGE
UNIVERSITY PRESS

CAMBRIDGE
UNIVERSITY PRESS

University Printing House, Cambridge CB2 8BS, United Kingdom

One Liberty Plaza, 20th Floor, New York, NY 10006, USA

477 Williamstown Road, Port Melbourne, VIC 3207, Australia

314-321, 3rd Floor, Plot 3, Splendor Forum, Jasola District Centre, New Delhi - 110025, India

103 Penang Road, #05-06/07, Visioncrest Commercial, Singapore 238467

Cambridge University Press is part of the University of Cambridge.

It furthers the University's mission by disseminating knowledge in the pursuit of
education, learning and research at the highest international levels of excellence.

www.cambridge.org
Information on this title: www.cambridge.org/9780521888608

© Cambridge University Press 2010

First published 2010

A catalogue record for this publication is available from the British Library

Library of Congress Cataloging in Publication data
Justesen, Jørn.
Two-dimensional information theory and coding : with application to graphics and high-density storage
media / Jørn Justesen and Søren Forchhammer.
    p.   cm.
Includes bibliographical references and index.
ISBN 978-0-521-88860-8 (hardback)
1. Coding theory.   2. Information storage and retrieval systems – Mathematics.   3. Bipartite graphs.
4. Computer storage devices – Mathematical models.   5. Computer graphics – Mathematics.   6. Data
visualization.   7. Binary system (Mathematics)   I. Forchhammer, Søren.   II. Title.
QA268.J87   2010
005.7′2 – dc22   2009020679

ISBN 978-0-521-88860-8 Hardback

Contents

Preface

Shannon's paper from 1948, which presented information theory in a way that already included most of the fundamental concepts, helped bring about a fundamental change in electronic communication. Today digital formats have almost entirely replaced earlier forms of signaling, and coding has become a universal feature of communication and data storage.

Information theory has developed into a sophisticated mathematical discipline, and at the same time it has almost disappeared from textbooks on digital communication and coding methods. This book is a result of the authors' desire to teach information theory to students of electrical engineering and computer science, and to demonstrate its continued relevance. We have also chosen to mix source coding and error-correcting codes, since both are components of the systems we focus on.

Early attempts to apply information-theory concepts to a broad range of subjects met with limited success. The development of the subject has mostly been fuelled by the advances in design of transmitters and receivers for digital transmission such as modem design and other related applications. However, more recently the extensive use of digitized graphics has made possible a vast range of applications, and we have chosen to draw most of our examples from this area.

The first five chapters of the book can be used for a one-semester course at the advanced-undergraduate or beginning-graduate level. Chapter 6 serves as a transition from the basic subjects to the more complex environments covered by current standards. The last chapters introduce problems related to two-dimensional storage and other current applications.

We would like to thank the graduate students and post-docs who have participated in our work and contributed to figures in the book.

Søren Forchhammer and Jørn Justesen

1 Introduction to information theory

1.1 Introduction

In this chapter we present some of the basic concepts of information theory. The situations we have in mind involve the exchange of information through transmission and storage media designed for that purpose. The information is represented in digital formats using symbols and alphabets taken in a very broad sense. The deliberate choice of the way information is represented, often referred to as coding, is an essential aspect of the theory, and for many results it will be assumed that the user is free to choose a suitable code.

We present the classical results of information theory, which were originally developed as a model of communication as it takes place over a telephone circuit or a similar connection. However, we shall pay particular attention to two-dimensional (2-D) applications, and many examples are chosen from these areas. A huge amount of information is exchanged in formats that are essentially 2-D, namely web-pages, graphic material, etc. Such forms of communication typically have an extremely complex structure. The term media is often used to indicate the structure of the message as well as the surrounding organization. Information theory is relevant for understanding the possibilities and limitation of many aspects of 2-D media and one should not expect to be able to model and analyze all aspects within a single approach.

1.2 Entropy of discrete sources

A *discrete information source* is a device or process that outputs symbols at discrete instances from some finite alphabet $A = \{x_1, x_2, \ldots, x_r\}$. We usually assume that a large number of such symbols will have to be processed, and that they may be produced at regular instances in time or read from various positions in the plane in some fashion. The use of a large number of symbols allows us to describe the frequencies in terms of a probability distribution.

A single symbol can be thought of as a random variable, X, with values from the finite alphabet A. We assume that all variables have the same probability distribution, and the probability of x_i is written as $P(x_i)$ or $P[x = x_i]$. It is often convenient to refer to the probability distribution, $P(X)$, of the variable as a vector of probabilities $\mathbf{p}_X$, where the values are listed in some fixed order.

As a measure of the amount of information provided by X we introduce the entropy of a random variable.

Definition. The *entropy* of the variable X is

$$H(X) = E[\log(1/P(X))] = -\sum P(x)\log P(x), \qquad (1.1)$$

where $E[\ ]$ indicates the expected value.

The word entropy (as well as the use of the letter H) has its historic origin in statistical mechanics. The choice of the base for the logarithms is a matter of convention, but we shall always use binary logarithms and refer to the amount of information as measured in *bits*.

We note that

$$0 \leq H(X) \leq \log r.$$

The entropy is always positive since each term is positive. The maximal value of $H(X)$ is reached when each value of X has probability $1/r$ and the minimal when one outcome has probability 1 (See Exercise 1.1). When $r = 2^m$, the variable provides at most m bits of information, which shows that the word bit in information theory is used in a way that differs slightly from common technical terminology. If X^N represents a string of N symbols, each with the same alphabet and probability distribution, it follows from (1.1) that the maximal entropy is NH, and that this maximum is reached when the variables are independent. We refer to a source for which the outputs are independent as a *memoryless source*, and if the probability distributions also are identical we describe it as i.i.d. (independent identically distributed). In this case we say that the entropy of the source is H bits/symbol.

Example 1.1 (Entropy of a binary source).   *For a binary memoryless source with probability distribution $P = (p, 1 - p)$ we get the entropy (1.1) as*

$$H(p) = -p \log p - (1 - p)\log(1 - p), \qquad (1.2)$$

where $H(p)$ is called the (binary) entropy function. The entropy function is symmetric with respect to $p = 1/2$, and reaches its maximal value, 1, there. For small p, H increases quickly with increasing p, and $H(0.11) = 0.5$.

We note that the same letter, H, is used to indicate the entropy of a random variable and the value of the entropy function for a particular distribution, $(p, 1 - p)$. We shall use the same notation $H(P)$ when P is a general discrete probability distribution.

Example 1.2 (Run-length coding).   *For a binary source with a small value of $P(1)$ it may be useful to segment a string of symbols into "runs" of 0s ending with a 1. In this way the source is converted into a memoryless source with alphabet $\{0, 1, 2, \ldots, n\}$, where the symbols indicate the length of the segment. The value k represents the string $0^k 1, k < n$. If runs of more than n 0s are possible, the value $k = n$ represents the string*

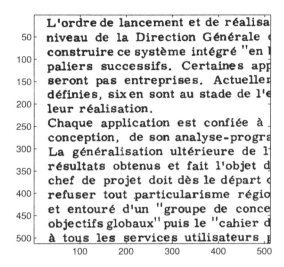

Figure 1.1. A binary image of a printed text.

0^n, *i.e. a run of 0s that is not terminated. Run-length coding may be used to code binary images of text along the rows (Fig. 1.1).*

For a pair of random variables, (X, Y), with values from finite alphabets $\mathcal{A}_x = \{x_1, x_2, \ldots, x_r\}$ and $\mathcal{A}_y = \{y_1, y_2, \ldots, y_s\}$, the conditional probability of y_j given x_i is $P(y_j|x_i) = q_{ij}$. It is often convenient to refer to the conditional probability distribution as the matrix $\mathbf{Q} = P(Y|X) = [q_{ij}]$. The probability distribution of the variable, Y, then becomes

$$\mathbf{p}_Y = \mathbf{p}_X \mathbf{Q},$$

where $\mathbf{p}_X$ and $\mathbf{p}_Y$ are the distributions of X and Y, respectively.

Definition. The *conditional entropy*

$$H(Y|X) = -\sum_{i,j} P(x_i, y_j)\log P(y_j|x_i)$$

$$= -\sum_i P(x_i) \sum_j P(y_j|x_i)\log P(y_j|x_i) \tag{1.3}$$

can be interpreted as the additional information provided by Y when X is known.

For any particular y_j, the average value of $P(y_j|x_i)\log P(y_j|x_i)$ is given by $-P(y_j)\log P(y_j)$. Since the function $-u \log u$ is concave (has negative second derivative), we can use Jensen's inequality, which states that for such a function

$$f(E(u)) \geq E[f(u)].$$

Thus, on applying Jensen's inequality with $u = p(y|x)$ to (1.3), it follows that the entropy is never increased by conditioning, i.e. $H(Y) \geq H(Y|X)$.

For a string of variables, $X_1, X_2, \ldots, X_n$ we have

$$P(X_1, X_2, \ldots, X_n)$$
$$= P(X_1)P(X_2|X_1)P(X_3|X_1, X_2) \ldots P(X_n|X_1, X_2, \ldots, X_{n-1})$$

and thus, taking the expectation of the logarithm,

$$H(X_1, X_2, \ldots, X_n)$$
$$= H(X_1) + H(X_2|X_1) + \cdots + H(X_n|X_1, X_2, \ldots, X_{n-1}). \tag{1.4}$$

This result is known as the *chain rule*, and expresses the total amount of information as a sum of information contributions from variable X_i when values of the previous variables are known. When the variables are not independent, the chain rule is often a valuable tool with which to decompose the probabilities and entropies of the entire string of data into terms that are easier to analyze and calculate.

Example 1.3 (Entropy by the chain rule). *If X takes values in $\mathcal{A} = \{x_1, x_2, x_3, x_4\}$ with probability distribution $(1/4, 1/4, 1/3, 1/6)$, we can of course calculate the entropy directly from the definition. As an alternative, we can think of the outcome as a result of first choosing a subset of two values, represented by the variable Y. We can then express the entropy by the chain rule:*

$$H(X) = H(Y) + H(X|Y) = H(1/2) + (1/2)H(1/2) + (1/2)H(1/3)$$
$$= 3/2 + (\log 3 - 2/3)/2 = 7/6 + (1/2)\log 3.$$

1.3   Mutual information and discrete memoryless channels

For a pair of random variables, (X, Y), with values from finite alphabets $\mathcal{A}_x = \{x_1, x_2, \ldots, x_r\}$ and $\mathcal{A}_y = \{y_1, y_2, \ldots, y_s\}$, the *mutual information* is defined as

$$I(Y; X) = \sum_{x,y} P(x, y)\log\left(\frac{P(y|x)}{P(y)}\right) = E\left[\log\left(\frac{P(y|x)}{P(y)}\right)\right]. \tag{1.5}$$

The mutual information is a measure of the amount of information about X represented by Y. It follows from the definition that the mutual information can be expressed in terms of entropies as

$$I(X; Y) = H(Y) - H(Y|X) = H(X) - H(X|Y). \tag{1.6}$$

The properties of the conditional entropy imply that the mutual information $I(X; Y) \geq 0$. It is 0 when X and Y are independent, and achieves the maximal value $H(X)$ when each value of X gives a unique value of Y.

It may also be noted that $I(Y; X) = I(X; Y)$. The symmetry follows from rewriting $P(y|x)$ as $P(y, x)/P(x)$ in the definition of I. This is an unexpected property since there is often a causal relation between X and Y, and it is not possible to interchange cause and effect.

From the chain rule for entropy we get

$$I(Y; X_1, X_2, \ldots, X_n)$$
$$= I(Y; X_1) + I(Y; X_2|X_1) + \cdots + I(Y; X_n|X_1, X_2, \ldots, X_{n-1}), \quad (1.7)$$

which we refer to as the *chain rule for mutual information.*

A discrete information channel is a model of communication where X and Y represent the input and output variables, and the channel is defined by the conditional probability matrix, $\mathbf{Q} = P(Y|X) = [q_{ij}]$, which in this context is also called the transition matrix of the channel. When the input distribution $P(X)$ is given, we find the output distribution as

$$\mathbf{p}_Y = \mathbf{p}_X \mathbf{Q}.$$

Example 1.4 (The binary symmetric channel). *The single most important channel model is the binary symmetric channel (BSC). This channel models a situation in which random errors occur with probability p in binary data. For equally distributed inputs, the output has the same distribution, and the mutual information may be found from (1.6) as*

$$C = I(Y; X) = 1 - H(p), \quad (1.8)$$

where H again indicates the binary entropy function. Thus $C = 1$ for $p = 0$ or 1, C has minimal value 0 for $p = 1/2$, and $C = 1/2$ for $p = 0.11$.

For two variables we can combine (1.4) and (1.6) to get

$$H(X, Y) = H(X) + H(Y) - I(X; Y).$$

Mutual information cannot be increased by data processing. If Z is a variable that is obtained by processing Y, but depends on X only through Y, we have from (1.7)

$$I(X; Z, Y) = I(X; Y) + I(X; Z|Y) = I(X; Z) + I(X; Y|Z).$$

Here $I(X; Z|Y) = 0$, since X and Z are independent given Y, and $I(X; Y|Z) \geq 0$. Thus

$$I(X; Z) \leq I(X; Y). \quad (1.9)$$

1.4 Source coding for discrete sources

In this section we demonstrate that the entropy of a message has a more operational significance: if the entropy of the source is H, a message in the form of a string of N symbols can, by suitable coding, be represented by about NH binary symbols. We refer to such a process as *source coding.*

1.4.1   The Kraft inequality

We shall prove the existence of efficient source codes by actually constructing some codes that are important in applications. However, getting to these results requires some intermediate steps.

A binary variable-length source code is described as a mapping from the source alphabet $\mathcal{A}$ to a set of finite strings, C from the binary code alphabet, which we always denote $\{0, 1\}$. Since we allow the strings in the code to have different lengths, it is important that we can carry out the reverse mapping in a unique way. A simple way of ensuring this property is to use a *prefix code*, a set of strings chosen in such a way that no string is also the beginning (prefix) of another string. Thus, when the current string belongs to C, we know that we have reached the end, and we can start processing the following symbols as a new code string. In Example 1.5 an example of a simple prefix code is given.

If c_i is a string in C and $l(c_i)$ its length in binary symbols, the *expected length* of the source code per source symbol is

$$L(C) = \sum_{i=1}^{N} P(c_i)l(c_i).$$

If the set of lengths of the code is $\{l(c_i)\}$, any prefix code must satisfy the following important condition, known as the *Kraft inequality*:

$$\sum_i 2^{-l(c_i)} \leq 1. \tag{1.10}$$

The code can be described as a binary search tree: starting from the root, two branches are labelled 0 and 1, and each node is either a leaf that corresponds to the end of a string, or a node that can be assumed to have two continuing branches. Let l_m be the maximal length of a string. If a string has length $l(c)$, it follows from the prefix condition that none of the $2^{l_m - l(c)}$ extensions of this string are in the code. Also, two extensions of different code strings are never equal, since this would violate the prefix condition. Thus by summing over all codewords we get

$$\sum_i 2^{l_m - l(c_i)} \leq 2^{l_m}$$

and the inequality follows. It may further be proven that any uniquely decodable code must satisfy (1.10) and that if this is the case there exists a prefix code with the same set of code lengths. Thus restriction to prefix codes imposes no loss in coding performance.

Example 1.5 (A simple code).   *The code* $\{0, 10, 110, 111\}$ *is a prefix code for an alphabet of four symbols. If the probability distribution of the source is* $(1/2, 1/4, 1/8, 1/8)$*, the average length of the code strings is* $1 \times 1/2 + 2 \times 1/4 + 3 \times 1/4 = 7/4$*, which is also the entropy of the source.*

If all the numbers $-\log P(c_i)$ were integers, we could choose these as the lengths $l(c_i)$. In this way the Kraft inequality would be satisfied with equality, and furthermore

$$L = \sum_i P(c_i)l(c_i) = -\sum_i P(c_i)\log P(c_i) = H(X)$$

and thus the expected code length would equal the entropy. Such a case is shown in Example 1.5. However, in general we have to select code strings that only approximate the optimal values. If we round $-\log P(c_i)$ to the nearest larger integer $\lceil -\log P(c_i) \rceil$, the lengths satisfy the Kraft inequality, and by summing we get an upper bound on the code lengths

$$l(c_i) = \lceil -\log P(c_i) \rceil \le -\log P(c_i) + 1. \tag{1.11}$$

The difference between the entropy and the average code length may be evaluated from

$$H(X) - L = \sum_i P(c_i)\left[\log\left(\frac{1}{P(c_i)}\right) - l_i\right] = \sum_i P(c_i)\log\left(\frac{2^{-l_i}}{P(c_i)}\right)$$

$$\le \log \sum_i 2^{-l_i} \le 0,$$

where the inequalities are those established by Jensen and Kraft, respectively. This gives

$$H(X) \le L \le H(X) + 1, \tag{1.12}$$

where the right-hand side is given by taking the average of (1.11).

The loss due to the integer rounding may give a disappointing result when the coding is done on single source symbols. However, if we apply the result to strings of N symbols, we find an expected code length of at most $NH + 1$, and the result per source symbol becomes at most $H + 1/N$. Thus, for sources with independent symbols, we can get an expected code length close to the entropy by encoding sufficiently long strings of source symbols.

Huffman coding

In this section we present a construction of a variable-length source code with minimal expected code length. This method, *Huffman coding*, is a commonly used coding technique. It is based on the following simple recursive procedure for binary codewords.

(1) Input: a list of symbols, S, and their probability distribution.
(2) Let the two smallest probabilities be p_i and p_j (maybe not unique).
(3) Assign these two symbols the same code string, α, followed by 0 and 1, respectively.
(4) Merge them into a single symbol, a_{ij}, with probability $p(a_{ij}) = p_i + p_j$.
(5) Repeat from Step 1 until only a single symbol is left.

The result of the procedure can be interpreted as a decision tree, where the code string is obtained as the labels on the branches that lead to a leaf representing the symbol. It

Table 1.1. Huffman coding (S denotes symbols and the two bits determined in each stage are indicated by italics)

Stage 1 (five symbols)			Stage 2 (four Symbols)			Stage 3 (three symbols)			Stage 4 (two symbols)		
Prob	S	C	Prob	S	C	Prob	S	C	Prob	S	C
0.50	a_1	0	0.50	a_1	0	0.50	a_1	0	0.50	a_1	0
0.17	a_2	100	0.19	a_{45}	11	0.31	a_{23}	10	0.50	a_{2-5}	1
0.14	a_3	101	0.17	a_2	100	0.19	a_{45}	11			
0.12	a_4	110	0.14	a_3	101						
0.07	a_5	111									

follows immediately from the description that the code is a prefix code. It may be noted that the algorithm constructs the tree by going from the leaves to the root.

To see that the Huffman code has minimal expected length, we first note that in any optimal prefix code there are two strings with maximal length differing in the last bit. These strings are code strings for the symbols with the smallest probabilities (Steps 1 and 2). If the symbols with the longest strings did not have minimal probabilities, we could switch the strings and get a smaller expected length. If the longest string did not have a companion that differs only in the last bit, we could eliminate the last bit, use this one-bit-shorter codeword and preserve the prefix property, thus reducing the expected length. The expected length can be expressed as

$$L = L' + p_i + p_j,$$

where L' is the expected length of the code after the two symbols have been merged, and the last terms were chosen to be minimal. Thus the reduced code must be chosen as an optimal code for the reduced alphabet with α (Step 3) and the same arguments may be repeated for this (Step 4). Thus, the steps of Huffman coding are consistent with the structure of an optimal prefix code. Further, the code is indeed optimal, insofar as restricting the class of codes to prefix codes gives no loss in coding performance, as previously remarked.

It follows from the optimality of the Huffman code that the expected code length satisfies (1.12).

An example of the Huffman code procedure is given in Table 1.1. For convenience the symbols are ordered by decreasing probabilities.

By extending the alphabet and code N symbols at a time, an average code length per original symbol arbitrarily close to the entropy may be achieved. The price is that the number of codewords increases exponentially in N.

1.5   Arithmetic codes

Arithmetic coding can code a sequence of symbols, x^n, using close to $-\log_2 p(x^n)$ bits, in accordance with the entropy, with a complexity which is only linear in n.

This is achieved by processing one source symbol at a time. Furthermore, arithmetic coding may immediately be combined with adaptive (dynamically changing) estimates of probabilities.

Arithmetic coding is often referred to as coding by interval subdivision. The coding specifies a number (pointing) within a subinterval of the unit interval $[0; 1[$, where the lower end point is closed and the upper end point is open. The codeword identifying the subinterval, and thereby the sequence that is coded, is a binary representation of this number.

1.5.1 Ideal arithmetic coding

Let x^n denote a sequence of symbols to be coded. $x^n = x_1 \ldots x_n$ has length n. Let x_i be the symbol at i. Let $P(x^n)$ denote a probability measure of x^n. At this point we disregard the restriction of finite precision of our computer, but this issue is discussed in the following sections. We shall associate the sequence of symbols x^n with an interval $[C(x^n); C(x^n) + P(x^n)[$. The lower bound $C(x^n)$ may be used as the codeword. This interval subdivision may be done sequentially, so we need only calculate the subinterval and thereby the codeword associated with the sequence we are coding.

If we were to order the intervals of all the possible sequences of length n one after another, we would have partitioned the unit interval $[0; 1[$ into intervals of lengths equal to the probabilities of the strings. (Summing the probabilities of all possible sequences adds to unity since we have presumed a probability measure.)

For the sake of simplicity we restrict ourselves to the case of a binary source alphabet. The sequence x^n is extended to either x^n0 or x^n1. The codeword subinterval (of x^{n+1}) is calculated by the following recursion:

$$C(x^n0) = C(x^n) \qquad \text{if } x_{n+1} = 0, \tag{1.13}$$

$$C(x^n1) = C(x^n) + P(x^n0) \qquad \text{if } x_{n+1} = 1, \tag{1.14}$$

$$P(x^ni) = P(x^n)P(i|x^n), \quad i = 0, 1. \tag{1.15}$$

The width of the interval associated with x^n equals its probability $P(x^n)$. We note that (1.15) is a decomposition of the probability leading to the chain rule (1.4). Using pointers with a spacing of powers of $1/2$, at least one pointer of length $\lceil -\log_2(P(x^n)) \rceil$ will point to the interval. ($\lceil y \rceil$ denotes the ceiling or round-up value of y.) This suggests that arithmetic coding yields a code length that on average is very close to the entropy.

Before formalizing this statement, the basic recursions of binary arithmetic coding (1.13)–(1.15) are generalized for a finite alphabet, $\mathcal{A}$:

$$C(x^n0) = C(x^n),$$

$$C(x^ni) = C(x^n) + \sum_{j<i} P(x^nj), \quad i = 1, \ldots, |\mathcal{A}| - 1,$$

$$P(x^ni) = P(x^n)P(i|x^n), \quad i = 0, \ldots, |\mathcal{A}| - 1,$$

where $|\mathcal{A}|$ is the size of the symbol alphabet. Compared with the binary version, we still partition the interval into two for the lower bound, namely $j < i$ and $j \geq i$. The last equation again reduces the interval to a size proportional to the probability of x^{n+1}, i.e. $P(x^{n+1})$. One difference is that the selected subinterval may have intervals both above and below.

Let a tag $T(x^n)$ denote the center point of the interval, $[C(x^n); C(x^n) + P(x^n)[$, associated with x^n by the recursions. We shall determine a precision of $l(x^n)$ bits, such that the codeword defined belongs to a prefix code for X^n. Now choose

$$l(x^n) = \lceil -\log_2(P(x^n)) \rceil + 1 < \log_2(P(x^n)) + 2. \tag{1.16}$$

Truncating the tag to $l(x^n)$ bits by (1.16) gives the codeword $c = \lfloor T(x^n) \rfloor_{l(x^n)}$, where $\lfloor z \rfloor_k$ denotes the truncation of a fractional binary number z to a precision of k digits.

Choosing $l(x^n)$ by (1.16) ensures not only that this value, c, points within the interval but also that $\lceil T(x^n) \rceil_{l(x^n)} (= \lfloor T(x^n) \rfloor_{l(x^n)} + 2^{-l(x^n)})$ points within the interval.

This means that the codeword c will not be the prefix of the codeword generated by the procedure for any other outcome of X^n. Since this codeword construction is prefix-free it is also uniquely decodable.

Taking the expectation of (1.16) gives the right-hand side of

$$H(X^n) < L < H(X^n) + 2.$$

Thus the expected code length per source symbol, L/n, will converge to the entropy as n approaches ∞.

In addition to the fact that the arithmetic coding procedure above converges to the entropy with a complexity linear in the length of the sequence, it may also be noted that the probability expression is general, such that conditional probabilities may be used for sources with memory and these conditional probabilities may vary over the sequence. These are desirable properties in many applications. The issue of estimating these conditional probabilities is addressed in Chapter 5.

1.5.2   Finite-precision arithmetic coding

A major drawback of the recursions is that, even with finite-precision conditional probabilities $P(i|j)$ in (1.15), the left-hand side will quickly need a precision beyond any register width we may choose for these calculations. (We remind ourselves that it is mandatory that the encoder and decoder can perform exactly the same calculations.) There are several remedies to this. The most straightforward is (returning to the binary case) to approximate the multiplication of (1.15) by

$$\bar{P}(x^n i) = \lfloor \bar{P}(x^n) \bar{P}(i|x^n) \rfloor_k, \quad i = 0, 1, \tag{1.17}$$

i.e. k represents the precision of the representation of the interval. The relative position (or exponent) of the k digits is determined by the representation of $C(x^n)$ as specified below.

Having solved the precision problem of the multiplication, we still face the problem of representing $C(x^n)$ with adequate precision and the related problem of avoiding having to code the whole data string before generating output bits. Adequate precision for unique decodability may be achieved by ensuring that all symbols at all points are associated with a non-empty interval. (To reduce the loss of coding efficiency we may desire more precision than this in some cases.) To make efficient use of the representation, consider the code interval $[C(x^n); C(x^n) + P(x^n)[$. Reading the binary fractional digits of the two interval end points from the left, the first, say, r bits are identical. These may be shifted out of the register since they will not change further in the recursion. This is the same as scaling $C(x^n)$ with 2^r. We need only consider the fractional part of the scaled number. $P(x^n)$ is scaled by the same number and the first q bits of the fractional part give the precision of (1.17). This has reduced the problem to what may be described as a carry problem. We may encounter an interval of $[0.01^s; 0.10^t[$, where 1^s and 0^t denote s and t consecutive 1s and 0s, respectively. Since the leading bits are different and therefore scaling is not possible, s and t could assume arbitrarily large values and exceed the size of the register, k. One solution to this problem is bit-stuffing. When $C(x^n) = 0.x \ldots x01^{k-1}$, where k is the precision of (1.17), a 0 is inserted to trap a possible carry. The 0 is destuffed at the decoder when $k - 1$ consecutive trailing 1s are encountered.

1.5.3   Decoding arithmetic codes

The decoding is performed by reversing the process. Consider having decoded the data string, x^n, up to $t - 1$ as $x^{t-1} = u$. Now the question is that of whether the next symbol x_t is 0 or 1. This is resolved by observing that

$$C(x^n) - C(x^{t-1}) \geq \bar{P}(u0) \Leftrightarrow x_t = 1, \tag{1.18}$$

which, in turn, can be decided by looking at a fixed number of the next bits of the code string and carrying out the subtraction which is the inverse of (1.14).

Determining when to terminate the decoding requires extra information because arithmetic coding is not a "one source symbol" to "one codeword" mapping. The source symbols may be organized in blocks of predefined size or, if not, a solution is to start with a header stating the number of source symbols coded.

Arithmetic coding may also be applied to alphabets of any finite size. One way based on binary arithmetic coding is to decompose the values of the alphabet in a binary tree. Another approach is to use general recursion. In the latter case, solving the practical problems of finite precision, carry overflow, and decodability, including stopping, is similar but more complicated than the binary version.

A fast version of arithmetic coding is given in Appendix A.

Example 1.6 (Huffman coding as a special case of arithmetic coding).   *If we have to code (independent) events having probabilities that are powers of 1/2, a Huffman code is optimal. In this case we may describe the Huffman code as a special case of arithmetic coding (or simplify the arithmetic code into a Huffman code). Consider the probabilities of Ex. 1.5, $p(a) = 1/2$, $p(b) = 1/4$, and $p(c) = p(d) = 1/8$, and the code $a \rightarrow$ "0," $b \rightarrow$ "10," $c \rightarrow$ "110," $d \rightarrow$ "111." Choosing the symbols in alphabetic order, the symbols are associated with the intervals (base-2 notation*

Table 1.2. Huffman coding by arithmetic coding
($C(x^n)$ is the codeword of x^n)

	$[C(x^n); C(x^n) + P(x^n)[$	$P(x^n)$
Initial	$[0; 1[$	1
$x_1 = a$	.0; .1	1/2
$x_2 = a$	.00; .01	1/4
$x_3 = b$	.0010; .0011	1/16
$x_4 = c$	.0010110; .0010111	1/128

in parentheses) as follows: a $[0; 1/2[$ ($[0.0; 0.1[$), b $[1/2; 3/4[$ ($[0.10; 0.11[$), c $[3/4; 7/8[$ ($[0.110; 0.111[$), and d $[7/8; 1[$ ($[0.111; 1.000[$) (Table 1.2). For each interval trailing 0s are added so that both ends of the interval have the same precision. It may be noted that, for the lower bound of the intervals as written, the fractional part coincides with the Huffman codeword and thus the truncation in (1.17) becomes void. After each source symbol we may, for this special case, shift the coded bits out of the register and rescale our interval to the unit interval. Here we also use our knowledge that we will not have carry overflow.

The Huffman codeword of $aabc$ is 0 0 10 110. (The spaces are not part of the code). We see that the Huffman codeword is identical to the last value, $C(x^4)$, which is the arithmetic codeword. Therefore, we may interpret the Huffman codeword as a binary fractional number indicating a subinterval specified by the arithmetic code construction. In this special case we could interchange Huffman and arithmetic encoding and/or decoding.

1.5.4   Universal codes

The structure of the source codes presented in this chapter depends on the probability distribution of the symbols. In many applications these probabilities are unknown, and the distribution may vary between users. In a *universal code* the symbol alphabet is known, but, for any probability distribution, the code achieves an expected code length (per symbol) that exceeds the entropy, H, by at most a given constant.

One solution is outlined below. The code frame represents a block of symbols and consists of a header, which lists the code strings or the approximate probabilities for the source symbols, and a body of encoded symbols using the basic source code. For a given alphabet and any given (small) redundancy, the necessary accuracy of the probabilities may be determined. Selecting a finite-precision representation of the probabilities gives a length of the header that is upper bounded by a constant, L_h. If the body of the code block includes N encoded symbols, it has length about NH. Thus the loss due to the header is at most L_h/N. By encoding sufficiently long sequences in each frame, we can reduce the loss to any desired acceptable level. The solution may readily be applied to i.i.d. sources, but is not restricted to these. Universal source codes are treated in more detail in Chapter 5.

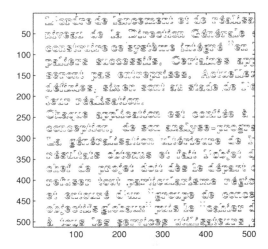

Figure 1.2. The vertical pixel differences, e_{ij}, of the image.

1.6 Entropy of images

An image with picture elements (or pixels), x_{ij}, where x_{ij} denotes the element at position (i, j), may be traversed, say, row by row to form a sequence. If the pixels are independent the entropy is expressed by (1.1). Normally the pixels are not independent, leading to a lower entropy. An example would be a binary image representing printed text and possibly graphics as line drawings, logos, halftones, etc. The text itself has a symbolic representation, namely the characters, but for printed text one or more fonts and font sizes may be chosen out of a very large set. Further, the graphics may have a very rich syntax. In representing such documents as binary images, the pixels are not independent, as illustrated below. Ways to model the dependency are treated in the next chapter, e.g. using the chain rule.

Example 1.7 (Binary images). *Figure 1.1 depicted a binary image, $x_{ij} \in \{0, 1\}$, showing a small part of a black-and-white representation of part of a document. The binary pixels display a high inter-pixel dependency. This may be illustrated e.g. by the simple operation of taking the difference $e_{ij} = (x_{ij} - x_{i-1j})$ mod 2 from the pixel above, which yields a less correlated representation (Fig. 1.2).*

Example 1.8 (Image bit-planes). *Natural images are often represented by 8 bits per pixel (bpp). A bit-plane is what we call the binary image obtained by taking a given bit of all pixels, thus forming a binary image. We refer to the least significant bit (LSB) of all pixels as the LSB bit-plane. For natural images these (LSB) bits are nearly independent (Fig. 1.4). This may be explained by invoking the noise inherent in the capture process.*

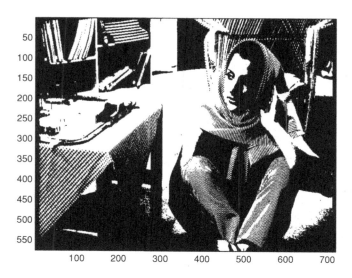

Figure 1.3. The most significant bit-plane of an 8-bpp image.

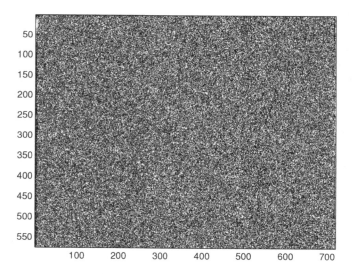

Figure 1.4. The least significant bit-plane of an 8-bpp image.

On the other hand, if we look at the most significant bit-plane, there is a high dependency, which is often seen as clear structures of the image (Fig. 1.3). Thus the entropy is smaller.

1.7   Notes

Information theory was first presented in the two-part paper [1]. The paper is not only of historical significance, but remains a readable introduction to the ideas of information and coding. The paper has been reprinted in various forms, but is more easily found as a PDF file online.

The popular textbook [2] has an emphasis on mathematical and statistical aspects.

A popular algorithm for arithmetic coding over a finite alphabet is given in [3].

Given the rapid advances in communication technology, earlier text books with a focus on applications to communication tend not to cover the methods of current standards.

Exercises

1.1 Show that, for a random variable, X, with values from an alphabet of size r, $0 \leq H(X) \leq r$. (Hint: use Jensen's inequality.)

1.2 A binary memoryless source has probability distribution $(P[0], P[1]) = (1/4, 3/4)$.
   (a) Find the entropy of the source.
   (b) Find the entropy of vectors of two and three symbols from the source.
   (c) Show that a sequence of symbols from the source can be uniquely segmented into strings from the set $\{1, 01, 001, 000\}$ (a method called run/length coding). What is the average length of the strings?
   (d) If the segmented sequence is considered as a memoryless source with four symbols, what is the entropy?

1.3 Consider a source with ten symbols and probabilities $i/45$ for $i = 1, 2, \ldots, 10$.
   (a) Find the entropy.
   (b) Construct a Huffman code for the source. What is the maximal length of the code strings?
   (c) What is the expected code length of the source? Compare the answer with the entropy.
   (d) Find sets of 10 strings with lengths from 2 to 4 such that the Kraft inequality is satisfied with equality.
   (e) Find a prefix code for the source with maximal string length 4 and minimal expected code length. Compare the expected length with that of the Huffman code.

1.4 For the same source as considered in Exercise 1.2, do the following.
   (a) Find Huffman codes for vectors of two and three symbols. Compare the expected length with the entropy of the source.
   (b) Find a Huffman code for the four strings from the run-length coding. Find the expected code length.
   (c) Compare the answer of the previous question with the entropy of the source and the average number of encoded source symbols.

1.5 For the source in Exercise 1.3, apply arithmetic coding to the sequence 10, 9, 8, 7, 10 by
   (a) calculating the mid-point tag and the necessary precision.
   (b) Decode on the basis of the finite-precision value of the tag.

1.6 For a binary sequence, perform a finite-precision arithmetic coding and decoding of the sequence.

References

[1] C. E. Shannon, "A mathematical theory of communication," *Bell System Techn. J.*, **27** (1948), 379–423 and 623–656.

[2] T. M. Cover and J. A. Thomas, *Elements of Information Theory* (Hoboken, NJ: Wiley, 2006).

[3] I. H. Witten, R. M. Neal, and J. G. Cleary, "Arithmetic coding for data compression," *Commun. ACM*, **30** (1987), 520–540.

2 Finite-state sources

2.1 Introduction

Typical data sources have complex structure, or we say that they exhibit memory. In this chapter we study some of the basic tools for describing sources with memory, and we extend the concept of entropy from the memoryless case discussed in Chapter 1.

Initially we describe the sources in terms of vectors or patterns that occur. Since the number of messages possible under a set of constraints is often much smaller than the total number of symbol combinations, the amount of information is significantly reduced. This point of view is reflected in the notion of combinatorial entropy. In addition to the structural constraints the sources can be characterized by probability distributions, and the probabilistic definition of entropy is extended to sources with memory.

We are particularly interested in models of two-dimensional (2-D) data, and some of the methods commonly used for one-dimensional (1-D) sources can be generalized to this case. However, the analysis of 2-D fields is in general much more complex. Information theory is relevant for understanding the possibilities and limitations of many aspects of 2-D media, but many problems are either intractable or even not computable.

2.2 Finite-state sources

The source memory is described by distinguishing several *states* that summarize the influence of the past. We consider only the cases in which a finite number of states is sufficient.

A *finite-state source* is specified by a finite set of states, $S = \{s_1, s_2, \ldots\}$, an *adjacency matrix*, $\mathbf{T}$, in which the element $t_{ij} = 1$ when a transition from state i to state j is possible and 0 otherwise, and an output U that associates a symbol from the finite alphabet $\mathcal{A}$ with each of the possible transitions.

A source is said to be *irreducible* if it is possible to reach every state in S from every other state in a finite number of transitions. Unless it specified otherwise, all sources are assumed to be irreducible. It is always possible to choose the states in such a way that, for all states s_i and output symbols u_{ij}, there is a unique next state s_j.

Strings of symbols that are produced by finite-state sources are sometimes called *finite-state languages*, suggesting that this is the simplest type of language structure. More elaborate models appear in definitions of programming languages and languages for describing graphic documents. Finite-state sources will be sufficient for our discussion of information-theory concepts.

In a string of symbols from a source, the variable associated with index t (often the time) is denoted $u(t)$.

2.2.1   Constrained sequences

Finite-state sources are used to describe sequences that satisfy various constraints. Some typical properties include the following.

- Run-length constraints. The 1-D (d, k)-*run-length-limited* (RLL) constraint consists of all binary words in which the run-lengths of 0s are between d and k, inclusive, except the first and the last runs, which may be shorter than d. Binary RLL (d, k) codes are used in some storage systems.
- Constraints on the sum of symbols in any finite sequence

$$Z_{min} < \sum_{t_1}^{t_2} u(t) < Z_{max}$$

for all t_1, t_2. The state can be characterized by the current value of the sum.
- A 1-D pattern constraint. N consecutive symbols in the sequence must belong to an admissible subset of $\mathcal{A}^N$ (or equivalently a set of specific patterns is excluded). States are characterized by $N - 1$ consecutive symbols.

2.2.2   Counting constrained sequences

A sequence, x with K symbols is denoted $x = x_1, x_2, \ldots, x_K = x_1^K = x^K$ using standard string notation. For a given finite-state source, let $F(n)$ be the number of admissible strings of length n. This number determines the number of messages of length n that we can distinguish.

We can express $F(n)$ in terms of the adjacency matrix, $\mathbf{T}$, which indicates the possible transitions between two states.

Let $\mathbf{u}$ denote the all-1s vector. Set $\mathbf{f}_0 = \mathbf{u}$. Let the vector $\mathbf{f}_n$ represent the number of configurations after n transitions for each of the symbols as final state. After n transitions, we have $\mathbf{f}_n = \mathbf{u}\mathbf{T}^n$, which thus expresses the number of sequences of length n having each of the final states. Summing over the states gives the total number of configurations of length n:

$$F(n) = \mathbf{u}\mathbf{T}^n\mathbf{u}',$$

where $\mathbf{u}'$ denotes the transpose of $\mathbf{u}$. With the string of n symbols we can code one out of $F(n)$ messages or $\lfloor \log_2 F(n) \rfloor$ bits.

We are particularly interested in an approximation to the number of long sequences.

Definition. The *combinatorial entropy*, C, of a finite state source is

$$C = \lim_{n \to \infty} \log F(n)/n, \qquad (2.1)$$

where the logarithm is to the base 2.

The combinatorial entropy expresses the number of bits per symbol we can code using long sequences. Asymptotically the largest eigenvalue, Λ, of $\mathbf{T}$ will dominate the expression, $F(n)$, and thus determine the growth rate of the number of configurations. (We note that the matrices are non-negative.) Assuming that $\mathbf{T}$ has s distinct eigenvalues, we can express $\mathbf{T}^n \mathbf{u} = \sum (\alpha_i \lambda_i^n)$, where λ_i are the eigenvalues of $\mathbf{T}$ and α_i are vectors given by the corresponding eigenvectors and $\mathbf{u}$. Thus, for large n, the largest eigenvalue, Λ, will dominate and the number of strings, $F(n)$, is approximately a constant times Λ^n and it follows immediately that

$$C = \log \Lambda. \qquad (2.2)$$

Example 2.1 (RLL $(1, \infty)$). *A simple example of a binary 1-D RLL constraint is defined by imposing a maximum run-length of 1 on the symbol 1, i.e. each $x_i = 1$ must be surrounded by 0s: $x_i = 1 \Rightarrow x_{i-1} = x_{i+1} = 0$. Thus there has to be a run of at least $d = 1$ 0 and there is no restriction on the runs of 0s between 1s. This is referred to as the RLL $(1, \infty)$ constraint.*

The length of the constraint is $N = 2$ and thus the states are composed of one binary symbol. The adjacency matrix, $\mathbf{T}$, is

$$\mathbf{T} = \begin{pmatrix} 1 & 1 \\ 1 & 0 \end{pmatrix},$$

reflecting that a transition leading to a 1 followed by a 1 is not admissible.

For the RLL $(1, \infty)$ constraint $\mathbf{f}_1 = \mathbf{u}\mathbf{T} = (2, 1)$. The largest eigenvalue of $\mathbf{T}$ is $\Lambda = (1 + \sqrt{5})/2$, which is the golden mean. The capacity is expressed by (2.2), $C = \log[(1 + \sqrt{5})] \approx 0.694$.

To count the precise number of admissible configurations of length n, $F(n)$, for the RLL $(1, \infty)$ constraint we may note that $F(n) = F(n-1) + F(n-2)$, $n > 2$ since any of the $F(n-1)$ admissible sequences of length $n-1$ may be concatenated with $x_n = 0$, whereas $x_n = 1$ implies that $x_{n-1} = 0$, which in turn may be concatenated with any of the $F(n-2)$ admissible sequences of length $n-2$. The recursion above gives the familiar Fibonacci sequence (starting with the third element) since $F(1) = 2$ and $F(2) = 3$.

2.3   Markov sources and Markov chains

To associate a probability distribution with strings from finite-state sources, we introduce the following important concept from probability theory. A *Markov chain* has the property that a symbol's dependence on the past is given by its dependence on the previous symbol:

$$p(x_t|x^{t-1}) = p(x_t|x_{t-1}). \tag{2.3}$$

Therefore, the joint probability of a sequence $x_1, x_2, \ldots, x_n = x_1^n$ may be written as

$$p(x_1^n) = p(x_1)p(x_2|x_1) \ldots p(x_t|x_{t-1}) \ldots p(x_n|x_{n-1}). \tag{2.4}$$

The variables are referred to as states and the transition probability for going from state i at time $t - 1$ to state j at time t is denoted $p_{ij} = P[x_t = j|x_{t-1} = i]$. These transition probabilities define the elements of the transition-probability matrix $\mathbf{P}$.

2.3.1   Stationary distribution

Given a probability distribution on the states, $\mathbf{p}_t$, at time t, the distribution at the next time instance, $t + 1$, is given by

$$\mathbf{p}_{t+1} = \mathbf{p}_t \mathbf{P}. \tag{2.5}$$

The *stationary distribution* on the states is a probability vector, $\mathbf{p}^*$, satisfying $\mathbf{p}^*\mathbf{P} = \mathbf{p}^*$.

A Markov chain is said to be *irreducible* if all states can be reached from a given initial state. As in the case of finite-state sources, we usually assume that this is the case. Sometimes it is convenient to also assume that the chain does not exhibit periodicity, i.e. there is not a period, τ, such that a certain subset of states can occur only for times $t = t' \bmod \tau$.

For an irreducible and aperiodic Markov chain the stationary probability distribution is the normalized unique left eigenvector of the transition matrix with eigenvalue 1. For any initial state, or any initial probability distribution on the states, the distribution chain will converge to the stationary distribution. Often the most practical way of computing the stationary distribution is to keep multiplying some distribution by the transition matrix until a stable result is obtained.

2.3.2   Reversal of a Markov chain

It is often useful to be able to read the symbols of a sequence in reverse order. In that context it is useful to note that any irreducible Markov chain can be reversed. To get the corresponding transition-probability matrix, proceed in the following way. First multiply the columns by the stationary distribution. Note that the entries in the resulting matrix are the probabilities of the individual transitions (they clearly sum to unity). Take the transpose of the matrix, and divide the columns by the stationary probabilities. It is clear that each transition occurs with the same probability as before, but, where it was expressed as $p_j^* p_{ij}$ before, it is now $p_i^* p_{ji}$, and it is readily checked that any sequence of symbols has the same probability as before.

2.3.3   Markov sources

In Markov chains the states are, by definition, the same as the previous symbols. In information theory it is convenient to have models that describe more general finite-state sources. We obtain such models by mapping Markov chains onto a different set of output symbols. A *Markov source* is a process whereby each transition in a Markov chain is mapped to a symbol from the source alphabet.

Where the state of a Markov chain is always obvious, a Markov source may produce the same output symbol when making transitions to different states. As noted earlier, the states of a finite-state source can be redefined to avoid this problem, but that is not the case when the probabilities of the transition are given. If a Markov source has the property that, for any given state and a given output symbol, there is a unique next state, the source is said to be *observable* (or unifilar). A source that cannot be brought into such a form is called a *hidden-state source*.

2.4 Entropy of Markov chains and Markov sources

2.4.1 Stochastic processes

A Markov chain is an example of the general concept of a (discrete) stochastic process, a probability assignment on strings from a given alphabet. Let X denote the process. Each variable in the string, X_i, is a stochastic variable.

To specify the properties of a string of arbitrary length, we need a way of assigning a probability to any string of finite length. Again we assume that such assignments are independent of the position in the string, i.e. a shift of a subsequence does not change the distribution. This is referred to as a *stationary process* and we consider only such processes.

Clearly the distributions for strings of different lengths must be consistent, but we do not need to discuss this in detail.

2.4.2 Entropy of Markov chains and Markov sources

The *entropy of a stochastic process*, $\{X\}$ is given by

$$H = \lim \frac{1}{n} H(X_1, X_2, \ldots, X_n). \tag{2.6}$$

For an irreducible Markov chain, using the chain rule, the entropy of the sequence with n elements is

$$H(X_1^n) = H(X_1) + H(X_2|X_1) + \cdots + H(X_n|H_{n-1}).$$

Taking the limits gives the per-symbol entropy for a stationary Markov chain,

$$H = \lim \frac{1}{n} \sum_{i=1}^{n} H(X_i|X_1^{i-1}) = H(X_j|X_{j-1}) = H(X_2|X_1), \tag{2.7}$$

where j is any positive integer, $j > 1$. In terms of the stationary probabilities and the transition matrix, the entropy is

$$H = \sum_i p_i^* \sum_j (-p_{ij} \log p_{ij}). \tag{2.8}$$

Example 2.2 (Run-length coding). *Run-length coding is used in some image-compression schemes, e.g. the early fax (facsimile) coding standard for coding binary images for transmission. The binary image is represented by the lengths of the alternating runs of consecutive black and white pixels. The observation is that the probability of long runs is higher than what a model of i.i.d. variables would assign. A simple model is given by a Markov chain. (See Exercise 2.2.)*

The entropy of an observable Markov chain is the same as the entropy of the underlying Markov chain, since there is a one-to-one mapping between the sequences. However, for a hidden-state source, the same output sequence may be produced by different state sequences, and the problem is more difficult. We give an upper and a lower bound for the entropy of a general Markov source.

The entropy is upper bounded as

$$H \leq H(X_j | X_{j-1}, X_{j-2}, \ldots, X_{j-m}) \tag{2.9}$$

and the entropy is lower bounded by

$$H \geq H(X_j | X_{j-1}, X_{j-2}, \ldots, X_{j-m+1}, S(j-m)), \tag{2.10}$$

where $S(t)$ is the state at time t. Both results follow from the observation that the exact entropy could be expressed by the conditional entropy given the entire past. However, if only a finite part is known, the correct value follows by conditioning on the remaining past, and the entropy would decrease. On the other hand, a known state is a stronger condition than the past, so the true entropy is greater than this value.

In principle we can find a close approximation to the entropy from these two bounds by letting m increase, but the convergence is slow.

A different approach to hidden Markov sources is based on describing the states of the source by a probability distribution on the states of the underlying Markov chain. If at time t this distribution, $s_h(t)$, is known, the distribution of the next state is found by multiplying by the transition matrix of the chain, (2.5). After the next output symbol, $u(t)$, is observed, we can eliminate all transitions that produce other symbols. If T_t indicates the remaining transition matrix, we can find the next distribution by normalizing the product:

$$s_h(t+1) = T_t s_h(t) / |T_t s_h(t)|.$$

In this way we convert the hidden-state source into an observable source, but in general there are infinitely many states.

2.5   Maximum-entropy Markov chains

Given a finite constraint and its adjacency matrix, **T**, a Markov chain is introduced by assigning probabilities to the state transitions. Let t_{ij} denote the element in the adjacency matrix representing whether a transition from state i to j is admissible. A probability measure may be induced by defining transition probabilities p_{ij} such that $p_{ij} \geq 0$ if $t_{ij} = 1$, $p_{ij} = 0$ if the transition is non-admissible ($t_{ij} = 0$), and $\sum_j p_{ij} = 1$. The output of the Markov chain will be admissible according to the constraint since all sequences containing a non-admissible configuration of length N will be assigned the probability 0.

The largest eigenvalue of the transition-probability matrix, $\mathbf{P}$, is 1, and the corresponding (right) normalized eigenvector is the stationary probability distribution $\mathbf{p}^*$ of the source.

For a given initial distribution $\mathbf{v}$, the distribution after n transitions be given by $\mathbf{v}\mathbf{P}^n$, which approaches the stationary distribution $\mathbf{p}^*$ with elements p_i^*. If this distribution is assumed from the start of the sequence, all variables, i will have the same distribution given by p_i^*. Thus the entropy of the source can be expressed by (2.8).

There exists a choice of transition probabilities, p_{ij}, such that the entropy of the Markov chain equals the combinatorial entropy given by the adjacency matrix. Let b_i be a left eigenvector of $\mathbf{T}$ with the largest eigenvector, λ, and choose the transition probabilities of the Markov chain as

$$p_{ij} = b_j/(b_i\lambda)$$

whenever $t_{ij} = 1$. It follows from the definition of the eigenvector that the transition probabilities sum to unity. Furthermore, all sequences of length n have approximately the same probability, λ^{-n}, which implies that the entropy is equal to the combinatorial entropy, which is the maximum entropy.

Example 2.3 (RLL $(1, \infty)$). *For the RLL $(1, \infty)$ constraint the largest eigenvalue of* $\mathbf{T}$ *is $\Lambda = (1 + \sqrt{5})/2$. Thus the transition matrix becomes*

$$\mathbf{P} = \begin{pmatrix} \Lambda^{-1} & \Lambda^{-2} \\ 1 & 0 \end{pmatrix}.$$

By writing the stationary distribution as $(\Lambda^2/(1 + \lambda^2), 1/(1 + \Lambda^2))$ we find the entropy to be

$$\frac{\Lambda^2}{1 + \Lambda^2} H(\Lambda^{-1}) = \log \Lambda,$$

thus achieving the value of the combinatorial entropy (2.2).

If the symbol distribution is known, we can consider a more limited maximization. Consider a finite-state source with given 0/1 adjacency matrix, $\mathbf{T}$. Let the states and the output be identified as in Markov chains, and let the stationary probability distribution, $\mathbf{p}^*$, on the symbols be given. We can then find the Markov chain with maximum entropy which is consistent with $\mathbf{T}$ and has a given stationary distribution, $\mathbf{p}^*$. First note that, if the columns of the transition matrix, $\mathbf{P}$, for a Markov chain are multiplied by the corresponding stationary probability, the rows also sum to the stationary distribution. We can find such a matrix in which each element represents the joint distribution of i and j by starting from the 0/1 adjacency matrix for the finite-state source and applying the procedure known as *iterative scaling*. From a given matrix we alternate between scaling the rows and scaling the columns to obtain desired values of their sums. In this example both the column and the row sums shall give the stationary probability, $\mathbf{p}^*$. If the matrix converges, we have the desired solution, otherwise the sum vectors are inconsistent with the initial matrix. We give a proof that the result has maximal entropy in Appendix B, which also has techniques for solving problems involving maximal entropies under other constraints on the symbols. A formal presentation of iterative scaling is given in Chapter 7.

2.6   Source coding for Markov chains

For each state, s_j, we consider the probability distribution of the next symbol, p_{ij}. This variable can be encoded using Huffman coding. The expected code length for that state satisfies (1.11). By combining (1.3) and (1.4) we see that this is true also for the combined expected length over all states. This approach implies having a table for each state of the Markov chain.

As in the memoryless case, we can get a more efficient code by combining the encoding of several symbols. Taking a fixed number of steps in the Markov chain rather than a single transition does not change the stationary distribution of the states. The expected code length per symbol approaches the entropy as $1/m$ when m transitions are encoded by a single code sequence.

Using arithmetic coding, the conditional probability in state s_j given by $P(x^n|x^{n-1}) = P(x^n|x_{n-1}) = p_{ij}$ may by used directly in the arithmetic recursion (1.15). For arithmetic coding the recursions are not influenced by the number of distinct states, but the distinct set of conditional probabilities will have to be handled, though.

2.7   Elementary two-dimensional sources

2.7.1   Structure in two-dimensional sources

The objective of information-theory models of 2-D sources is the description of graphics pages (printed or displayed on screens) and of constraints for 2-D storage media. Graphics pages can have quite complicated structure, and many of the formats that are used in current applications include tools for describing the layout of pages with mixed content, segments of symbols from large alphabets arranged in lines and with variable spacing between lines and within lines, etc. Some techniques for storage media are based on patterns that satisfy 2-D constraints. An information-theoretic analysis is possible only in very simple cases.

Example 2.4 (Crossword puzzles).   *Given a finite-state description of admissible strings, we can consider filling out an array in such a way that all rows and columns are admissible. Crossword puzzles (with various definitions of admissible words and phrases)*

*and several types of puzzles with approximately square pieces (subject to some require-
ment that adjacent edges match) can be described in this way. Such constructions
are called puzzles when the number of solutions is small. If the number of solutions
increases sufficiently quickly with the size, we can think of the array as an information source. Let the alphabet be* $[c, o, d, e]$ *with adjacency matrix, for both rows and
columns,*

$$\mathbf{T} = \begin{pmatrix} 1 & 0 & 1 & 0 \\ 1 & 1 & 1 & 1 \\ 1 & 0 & 1 & 0 \\ 1 & 1 & 1 & 1 \end{pmatrix}.$$

*Thus two consonants cannot be adjacent in any row or column. We do not consider
particular restrictions along the borders. Thus a valid array could be*

$$\begin{pmatrix} c & o & d & e \\ e & d & e & o \\ o & e & o & d \\ c & e & d & e \end{pmatrix}.$$

Example 2.5 (The hard-square constraint). *The 1-D RLL* $(1, \infty)$ *constraint may be
generalized to 2-D by requiring that it is satisfied for any row and any column of the
2-D array, i.e. in any row or column we cannot have a run of two consecutive 1s. This
may also be expressed as forbidding configurations having two neighboring 1s, i.e.
$x_{ij}x_{i-1,j} \neq 1$ and $x_{ij}x_{i,j-1} \neq 1$, where the elements are taken from the binary alphabet,
$x \in \{0, 1\}$. This constraint is called the hard-square constraint, where the 1s are the
squares and hard means that they are too big and hard to be neighbors (see Fig. 2.1).
The constraint is also known as the 2-D RLL* $(1, \infty)$ *constraint.*

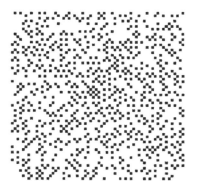

Figure 2.1. A hard-square configuration ([3], ©1999 IEEE).

2.7.2   Bands of finite height

Consider an M by N array of symbols, where initially we let M be a fixed small number, whereas N is much larger and can be chosen arbitrarily. We refer to such an array as a *band* of height M. Such bands can serve as an approximation to actual 2-D arrays, but it should be noted that there are also applications where such bands are appropriate models. These include 2-D printed barcodes obtained by stacking 1-D codes and lines of text. Another example is multi-track storage systems where parallel tracks are used to obtain higher densities in storage media. On a magnetic tape, a fixed (small) number of tracks may be written. Let s_i indicate a column of M symbols from the alphabet $\mathcal{A}$. Such a combination will serve as a state in a Markov chain describing the band, and the states may be any subset of $\mathcal{A}^M$. If there is a rule specifying which columns can be neighbors, we can describe the band as a finite-state source, and the combinatorial entropy can be defined as in Section 2.2.2. Thus the bands are a sequence of M-dimensional vectors. The band can be described by an adjacency matrix, as also discussed in Section 2.2.2. We can count the number of configurations and find the stationary distribution and entropy of the corresponding Markov chain. Such calculations are feasible even for very large matrices, but, since the number of states increases exponentially with M, the band can have only a very limited height.

Example 2.6 (A hard-square band).   *For a fixed height, M, the states of the band source are given by one column of M elements. For the 2-D RLL $(1, \infty)$ constraint, the number of states equals the number of admissible configurations of the 1-D RLL $(1, \infty)$ of length M and this number was shown to increase as the Fibonacci numbers in M (Example 2.1). For the height $M = 2$, the states are 00, 01, and 10, and the adjacency matrix is*

$$\mathbf{T}_2 = \begin{pmatrix} 111 \\ 101 \\ 110 \end{pmatrix},$$

reflecting that a transition from 01 or 10 to itself leads to two horizontally neighboring 1s, which is not admissible. The size of the transition matrix increases exponentially, so even for modest values of the height of the band it becomes too large for practical computations.

The entropy of the band is given by the largest eigenvalue of the adjacency matrix. The combinatorial entropy of a 2-D constraint (per symbol) is upper bounded by the entropy of the band since concatenating bands to cover the plane can only lead to restrictions. This will be discussed in more detail in Chapter 7.

2.7.3   Causal models

We can model information in 2-D as a finite M by N array of random variables with the same alphabet, $\mathcal{A}$. In the basic version of such a model we assume that all variables have the same probability distribution. It is an essential property of 2-D representations

Figure 2.2. A two-pixel neighborhood given by the causal neighbors of a row-by-row scan.

that there is no given ordering of the variables, and in general they may be processed in whatever order is convenient to the user. Clearly the order is of no importance if the variables are independent, and in that case the entropy of the source is the same as the entropy of a single variable.

We may decide to index the variables of an array using a single index. Usually the variables are processed row by row from the top, each row read from left to right. This is the order adopted in writing (in many languages) and in scanning documents for storage or transmission. The word causal refers to a model based on such an indexing, and it refers to the way the (artificial) indexing separates the array of symbols around the one currently being processed into a "past" (above and to the left) and a "future" (to the right and below).

For any indexing of the variables we can use the chain rule to express the entropy of the array as

$$H(X_1, X_2, \ldots, X_n)$$
$$= H(X_1) + H(X_2|X_1) + \cdots + H(X_n|X_1, X_2, \ldots, X_{n-1}).$$

If we make the additional assumption that each variable depends on the past only through variables "close" to it in the 2-D array, and that the size of the array is much larger than the size of this neighborhood (context), most variables (those not close to a border) will contribute the same amount to the entropy, i.e. the entropy of a single variable conditioned on the context, Fig. 2.2.

If the array is indexed X_{ij} and is processed row by row, we may consider the case in which the probability of row j' depends on the "past," $j < j'$, only through the previous row, $j' - 1$. Let X_p^{ij} denote the past elements prior to X_{ij}, i.e. $\{X_{i'j'}|X_{ij}, i = i' - 1 \vee i = i', j \leq j' - 1\}$. If the two consecutive rows can be described by a single Markov chain with pairs of symbols as states, each row is described by a Markov source that has the same states, but only one of the symbols as output. The two-row Markov chain has at most $\mathcal{A}^2$ states. If the two rows have the same distribution, we have a stationary causal model, in which each row is described conditioned on the one above by the Markov chain. Thus the entropy of the field is

$$H(X) = H(X_{ij}|X_p^{ij}) = H(X_i|X_{i-1}) = H(X_i, X_{i-1}) - H(X_i),$$

where a single index indicates entropies of row processes and $H(X_i)$ expresses the per-symbol entropy of the row process. In this way we have expressed the entropy of the field, but it should be noted that the Markov source for a single row is in general a function of a Markov chain, i.e. a hidden-state source, and it is not so simple to find the exact value of its entropy, $H(X_i)$, but it may be bounded from above, (2.9), and below, (2.10).

Example 2.7 (A two-row Markov chain for the hard-square model). *A two-row Markov chain for the hard-square model can easily be described as a chain with three states (see Ex. 2.6) and transition-probability matrix*

$$\mathbf{P}_2 = \begin{pmatrix} (1-2q) & q & q \\ (1-r) & 0 & r \\ (1-r) & r & 0 \end{pmatrix}.$$

It follows from the symmetry of the chain that the two rows have the same probability distribution, and the entropy of two rows can be calculated as described earlier. The chain can be used to specify a given row conditioned on the previous row, but in general the description of a row cannot be simplified from three to two states, and the entropy of the three-state source is not readily calculated. Calculations show that no choice of the two parameters gives a causal model, which we consider next.

2.8   Pickard random fields

We consider 2-D probability assignments that share the property that the symbols in rows and columns are outcomes of the same irreducible Markov chain over a finite alphabet $\mathcal{A}$.

As in the previous section, we describe the field by the distribution of two rows, and then derive a causal model in which a variable in the lower row is defined conditioned on the past. This can be repeated, describing the image row by row.

2.8.1   Independence assumptions

A state in the two-row Markov chain is defined by a pair of symbols, and the probability distribution of this pair, $X_{i,j}, X_{i+1,j}$, is found from the Markov chain describing a single column, where i and $i+1$ denote the upper and the lower row, respectively. The conditional probability of the next symbol in the upper row, $P[X_{i,j+1}|X_{i,j}]$, also follows from the single-row Markov chain, but we shall make the important assumption that it is independent of $X_{i+1,j}$,

$$P[X_{i,j+1}|X_{i,j}, X_{i+1,j}] = P[X_{i,j+1}|X_{i,j}]. \tag{2.11}$$

With this assumption, we can continue to find the conditional probabilities of the following symbols in the upper row, and the row simply becomes a Markov chain. The challenge is now to obtain the same distribution on the lower row.

There are two ways of assuring that the lower row is described by the same Markov chain. One is the condition

$$P[X_{i+1,j+1}|X_{i,j}, X_{i+1,j}] = P[X_{i+1,j+1}|X_{i+1,j}], \tag{2.12}$$

which is symmetric to the first condition and lets the lower row be continued forward in the same way. The alternative is

$$P[X_{i+1,j}|X_{i,j+1}, X_{i+1,j+1}] = P[X_{i+1,j}|X_{i+1,j+1}], \tag{2.13}$$

which allows the lower row to be continued in the reverse direction using the reversed Markov chain.

Two-dimensional fields based on these independence assumptions, namely (2.11) and either (2.12) or (2.13), are known as *Pickard random fields* (PRFs). Thus, for a Pickard random field, each row and each column is described by the Markov chain (rows from left to right and columns bottom up; the Markov chain may be different when reversed). We first consider the second version, (2.13).

2.8.2   Probability on a rectangle

Consider a rectangle of variables, $\{X_{i,j}|1 \le i \le I, 1 \le j \le J\}$, and a given two-row Markov chain. The independence condition (2.11) implies that the first row becomes a Markov chain and also that

$$P[X_{i,j}, X_{i,j+1}, X_{i+1,j}] = P[X_{i,j}]P[X_{i,j+1}|X_{i,j}]P[X_{i+1,j}|X_{i,j}].$$

This in turn means that also the first column may be described as a Markov chain. Having the first column and the first row, (2.11) implies that the probabilities of the remaining parts of the interior may be expressed row by row on the basis of the probability $P[X_{i+1,j+1}|X_{i,j}, X_{i+1,j}, X_{i+1,j}]$ derived from the two-row Markov chain. This 2-D version of the chain rule leads to the following expression for the probability of the variables of a $I \times J$ rectangle:

$$P[X_{1,1}] \prod_{j=1}^{J-1} P[X_{1,j+1}|X_{1,j}] \prod_{i=1}^{I-1} P[X_{i+1,1}|X_{i,1}]$$

$$\times \prod_{i=1}^{I-1} \prod_{j=1}^{J-1} P[X_{i+1,j+1}|X_{i,j}, X_{i+1,j}, X_{i+1,j}].$$

The expression is based on (2.11). Adding an extra independence condition (2.12) or (2.13) is, as noted, necessary in order for the field to be stationary.

2.8.3   Constructing a two-dimensional random field from a Markov chain

For ease of notation, the variables A, B, C, D are introduced to represent the four variables within a 2 by 2 square located at (i, j):

$$\begin{bmatrix} X_{ij} & X_{i,j+1} \\ X_{i+1,j} & X_{i+1,j+1} \end{bmatrix} = \begin{bmatrix} A & B \\ C & D \end{bmatrix}. \tag{2.14}$$

The first step is to define a distribution on $ABCD$. The next step is to argue that this leads to a stationary distribution such that the distribution is identical to $P(ABCD)$ for any 2 by 2 square, i.e. the same for any choice of i and j.

The distribution on ABC and that on BCD, respectively, are given by two transitions of the same Markov chain, e.g. starting with C: $C \rightarrow A \rightarrow B$ and $C \rightarrow D \rightarrow B$, which gives, using (2.11),

$$P(ABC) = P(C)P(A|C)P(B|A) = P(A)P(B|A)P(C|A), \qquad (2.15)$$

where the distributions $P(A)$ and $P(C)$ are identical and given by the stationary distribution of the Markov chain. We note that B and C are independent given A, i.e. $P[B|AC] = P[B|A]$. Likewise (2.13) may be expressed by

$$P(BCD) = P(C)P(D|C)P(B|D). \qquad (2.16)$$

Since the two distributions are equal, they have the same marginal distribution on BC. Deriving $P(D|ABC) = P(D|BC)$ from $P(BCD)$ is consistent and ensures stationarity since the distribution on each of the elements A, B, C, D is given by the same stationary Markov chain and likewise the distributions on neighboring pairs AB, CD, CA, DB are given by the same stationary Markov chain. This gives

$$P(D|BC) = P(BCD)/P(BC), \qquad (2.17)$$

$$P(BC) = \sum_d P(BC, D = d), \qquad (2.18)$$

where $P(BCD)$ is given by (2.16). This simple Pickard random field is called a two-pixel PRF.

To show that we can extend the description to a stationary field, consider two rows described by the Markov chain given by the conditional distribution $P(BD|AC)$.

Since $P[B|AC] = P[B|A]$, (2.15), row i becomes a Markov chain, since the conditional probability of each variable now depends only on the symbol to the left, and we can specify the distribution on the rest of row i by repeating application of this rule. Similarly we can ensure that row $i + 1$ is a Markov chain by requiring $P[C|BD] = P[C|D]$, (2.16), and repeating the argument but in the reverse direction using the reversed Markov chain. By construction the two Markov chains are the same when read in the same direction. So describing the second row conditional on the first row yields the same Markov description. By induction we can show that describing row $i + 1$ conditioned on row i yields the stationary Markov chain and thus a stationary field for all i, j. By the same line of argumentation, it may be shown that all columns are Markov chains. Since $P[C|AB] = P[C|A]$, (2.15), row j is a Markov chain. Considering a two-column process on columns j and $j + 1$, since $P[B|CD] = P[B|D]$, (2.16), column $j + 1$ is a Markov chain seen from the bottom up, but again, on considering the two Markov chains in the same direction, they again yield the same Markov description. By induction we get that all rows and all columns are Markov chains.

Owing to the stationarity, $H(D|ABC)$ will express the conditional entropy of elements not within the first row or column. For large rectangles, the conditional entropy, $H(D|ABC)$, will dominate. Furthermore, the entropy per symbol within the first row or column is not less than $H(D|ABC)$. For the two-pixel PRF, the conditional entropy becomes

$$H(D|ABC) = H(D|BC) = \sum_{bcd} P(BCD)\log\left(\frac{P(BC)}{P(BCD)}\right) \tag{2.19}$$

and thus is expressed directly in terms of the underlying 1-D Markov chain by (2.16)–(2.18).

One might suspect that the two independence conditions simply produced mutually independent rows, but that is not at all the case. In the next example a two-pixel PRF is defined such that the hard-square constraint is satisfied.

Example 2.8 (A Pickard random field for the hard-square constraint). *In Example 2.5, we considered the hard-square constraint on a binary field, i.e. the four immediate neighbors of a 1 must be zeros. Assume that rows and columns of the field are Markov chains with transition-probability matrix*

$$\mathbf{P} = \begin{pmatrix} 3/4 & 1/4 \\ 1 & 0 \end{pmatrix}.$$

Here the stationary distribution of the Markov chain is $(P[0], P[1]) = (4/5, 1/5)$. Let $A, B, C,$ and D denote the elements on a 2 by 2 square. In a causal model, $D = 1$ is possible only if the two symbols above (B) and to the left (C) are both zeros. This happens if the symbol in the diagonal (A) is a 1, and the probability of this event is $P(A = 1) = 1/5$. For a PRF, the probability of three zeros (ABC) in the 2 by 2 square can be calculated from the Markov chain by (2.15):

$$P[ABC = 000] = P[A = 0]P[B = 0|A = 0]P[C = 0|A = 0]$$
$$= P[0]P[0|0]^2 = 4/5(3/4)^2 = 9/20$$

(we may note that the reverse Markov chain is the same). Thus the total probability of the two neighbors (BC) being zeros is $1/5 + 9/20 = 13/20$, and, for the causal model to produce a 1 with probability 1/5, the conditional probability (2.17) must be $P(D = 1|B = C = 0) = 4/13$. Since $B = C = 0$ is the only case for which D is not determined beforehand (i.e. a forced 0), and thus contributes to the entropy, the conditional entropy is given by $H(D|ABC) = H(D|BC) = P[B = C = 0]H(D|B = C = 0) = 13/20 \times H(4/13) = 0.5788$, where $H(p)$ is the binary entropy function. Before it was argued in the general case that this is also the entropy of the field. (We would get the maximum entropy, 0.5831, for a PRF with Markov chains in the rows and columns for the hard-square constraint by choosing $P[1] = 0.2834$.)

The construction considered in this subsection usually gives an asymmetric distribution. Thus in the above example $P(CAB = 101) = 1/20$, but $P(ABD = 101) = 4/65$. We can modify the distribution by letting $P(D|ABC)$ depend on A, but in that case the entropy is reduced. Such a modification is considered in the following subsection.

The solution for the two-pixel PRF is simple, if we can accept the conditional independence relation, $P(D|ABC) = P(D|BC)$. For the general PRF, we have to specify

$X_{i+1,j+1}$ conditioned on the two symbols above, AB, and the one to the left, C, in such a way that the distribution is consistent with the distributions on three symbols that we have already defined and a causal definition of the field is obtained. The consistency requirement is given by

$$\sum_a (P(D|A = a, BC)P(A = a|BC)) = P(D|BC).$$

We usually have some freedom to choose this distribution. As for the simple case, the independence properties of the PRF will ensure that both rows in the two-row Markov chain are Markov chains themselves and specifying $P(ABC)$ and $P(BCD)$ by the same Markov chain again ensures stationarity.

2.8.4   Symmetric Pickard random fields

Pickard random fields with a more symmetric distribution can in some cases be found when the two-row Markov chain is symmetric. We can then write the second Pickard condition (2.12) as

$$P[D|AC] = P[D|C]. \tag{2.20}$$

It now follows that $P[CAB] = P[ACD]$, and we get a symmetric distribution by requiring that these distributions are the same as $P[ABD]$ and $P[CDB]$. For any distribution $P[ABCD]$ compatible with these conditions we can find the conditional distribution $P[D|ABC]$ and construct the field. A binary symmetric PRF for which the backward Markov chain is identical to the forward chain has three free parameters, e.g. $P[A = 1]$, $P[B = 1|A = 1]$, and $P[D = 1|A = 1, B = 1, C = 1]$.

For a Pickard model of an image, the conditional probabilities are given by $P(D|ABC)$, except for the first row and the first column. The sequence of conditional probabilities may sequentially be coded by arithmetic coding.

Example 2.9 (A symmetric PRF for the hard-square model). *As noted above, $P[BC = 11] = 1/20$, and we can choose the same value for $P[AD = 11]$. This implies that the conditional probability $P[D = 1|A = 1] = 1/4$. To get the correct total value of $P[D]$, we must have $P[D = 1|ABC = 000] = 1/3$. In this case we get slightly lower entropy,*

$$H(X) = 1/5 \times H(1/4) + 9/20 \times H(1/3) = 0.5755,$$

where $H(p)$ denotes the binary entropy function. This value may be compared to the value of 0.5788 in Example 2.8.

Example 2.10 (Covering the plane with colored rectangles). *Let the alphabet have three or more symbols (colors) and consider the constraint that, if three of the symbols in a 2 by 2 square have the same color, the last must also have that color. The constraint has the effect of separating the plane into rectangles in such a way that no two rectangles of*

Figure 2.3. Tiling the plane with rectangles of three different colors ([3], ©1999 IEEE).

Figure 2.4. A simple MRF neighborhood divided into the causal and non-causal parts of a row-by-row scan of the field.

the same color are neighbors (unless they can be merged into a larger rectangle). Since the constraint does not directly restrict the distribution of three symbols, fields with this property can satisfy the Pickard condition, and they may be constructed as described in this section. See Fig. 2.3.

2.9 Markov random fields

It is possible to avoid the extra assumption of the Pickard fields, and thus to get a generalization of Markov chains to two dimensions. A *Markov random field* (MRF) is a probability distribution on 2-D arrays of symbols such that the conditional distribution of any finite subset of the array depends only on the symbols bordering on the subset, i.e. the interior of the subset is independent of the exterior given the border of the subset (Fig. 2.4). We consider only a special situation, but it illustrates some of the difficulties with this concept.

Assume that the conditional distribution on a finite rectangular array depends on the rest of the field only through the variables on the rows and columns on either side of the array. For some constraint on the symbols, we assume that all permitted outcomes on the array have the same probability. It is easy to verify that this rule gives consistent definitions of distributions on subsets of the array. If it is possible to let the size of the array increase and find a limiting probability distribution, it is intuitively clear that we would have maximized the entropy, and that it would actually equal the combinatorial entropy for the constraint.

The problem with this approach is that it is difficult to count the configurations, and for this reason finding the actual value of the probabilities is not practical. Also there is no simple way of finding a limiting distribution for a large array, or for that matter of ensuring that the limit exists.

In most cases we could get a larger entropy with an MRF than with a PRF based on the same constraints. Thus we cannot in general reach the combinatorial entropy with a Pickard field.

2.10   Notes

The description of sources by Markov chains was already discussed in Shannon's paper. This topic is also treated in detail in the text by Cover and Joy (both references are given in Chapter 1). Pickard random fields are usually not covered in introductions to information theory or probability theory. The binary symmetric PRF was introduced in [1] and later generalized [2]. Constraints in 2-D such as the hard-square constraint and tiling with three colors were treated in [3]. We give more detail in Chapter 7.

Exercises

2.1 A Markov chain with four states has the following transition matrix:

$$M = \begin{bmatrix} 1/2 & 1/2 & 0 & 0 \\ 1/4 & 1/2 & 1/4 & 0 \\ 0 & 1/4 & 1/2 & 1/4 \\ 0 & 0 & 1/2 & 1/2 \end{bmatrix}.$$

(a) Find the stationary distribution and the entropy of the chain.

(b) M can describe a Markov source with alphabet $0, 1, -1$, where 0 is produced when the source remains in the same state, 1 when it moves up, and -1 when it moves down. What are the possible values of a long string of symbols?

(c) If the set of state transitions is unchanged, but the probabilities may be varied, what is the maximum entropy of any source with these states and transitions?

(d) What is the maximum entropy if the stationary distribution is fixed?

(e) Change the symbol alphabet to 1 in the two upper states, 0 in the lower states. Show that the state cannot always be determined from the output.

(f) Find upper and lower bounds on the entropy of the source.

(g) Improve the bounds by noting that the state of the source is known after any change of outputs from 0 to 1 or 1 to 0.

2.2 Run-length coding of the output of a Markov chain with two states 0 and 1, representing white and black, respectively, shall be analyzed. The Markov chain is given by $P(1|1) = 1/2$ and $P(0|0) = 7/8$.

(a) Calculate the stationary distribution and the entropy of the Markov chain.

(b) The black and white runs will alternate. Determine two Huffman codes, one for coding white runs and one for coding black runs. (For the white runs you introduce an escape character for runs longer than, say, 5. After the escape character, the same run-length code is used again.)

(c) Calculate the average code length for the run-length coding in (b).

2.3 A Markov chain with alphabet $\{a, b, c\}$ has transition probabilities

$$M = \begin{bmatrix} 1/2 & 1/2 & 0 \\ 0 & 1/2 & 1/2 \\ 1/2 & 0 & 1/2 \end{bmatrix}.$$

(a) Find the probabilities of all sequences of three symbols.

(b) Using this chain to describe rows and columns, we want to construct a two-pixel Pickard field. Find the conditional distribution $P[D|BC]$.

(c) Find the entropy of the field.

2.4 A binary Markov chain has transition probabilities

$$M = \begin{bmatrix} 1/4 & 3/4 \\ 3/4 & 1/4 \end{bmatrix}.$$

(a) Find the probabilities of all sequences of three symbols.

(b) Using this chain to describe rows and columns, we want to construct a Pickard field such that no 2 by 2 block contains four 0s or four 1s.

(c) Assume that we maintain the symmetry with respect to 0 and 1. Find the probabilities of all 2 by 2 blocks (introduce a parameter if the probabilities are not unique).

(d) Find the entropy of the field.

(e) Generate a graphic representation of some 16 by 16 outcomes (try various parameters in the underlying Markov chain).

References

[1] D. K. Pickard, "A curious binary lattice process," *J. Appl. Prob.*, **14** (1977), 717–731.

[2] D. K. Pickard, "Unilateral Markov fields," *Adv. Appl. Prob.*, **12** (1980), 655–671.

[3] S. Forchhammer and J. Justesen, "Entropy bounds for constrained two-dimensional random fields," *IEEE Trans. Inform. Theory*, **45** (1999), 118–127.

3 Channels and linear codes

3.1 Introduction

In Chapter 1 we briefly introduced an *information channel* as a model of a communication link or a related system where the input is a message and the output is an imperfect reproduction of it. In particular we also use this concept as a model of a storage system where input and output are separated in time rather than in space. In our presentation we do not refer to the underlying physical medium or discuss whether it is fundamentally continuous or quantized. The process of transmitting and receiving (writing and reading) is assumed to use finite alphabets, which may well be different, and it is understood that these alphabets represent a digital implementation of processes that make efficient use of the physical medium under the current technological and economic constraints. In this chapter we introduce the fundamentally important concept of channel capacity. It is defined in a straightforward way as the maximum of mutual information; however, the significance becomes clear only as we show how this is actually the amount of information that can be reliably transmitted through the channel. Reliable communication at rates approaching capacity requires the use of coding. For this reason we have chosen to present the basic concepts of channel coding in the same chapter and to emphasize the relation between codes and the information-theoretic quantities. In reality the codes that are used are matched to a few special channels, and other real channels are converted to or approximated by one of these types. Thus our discussion of capacity is related to such specific channels rather than a more general class.

3.2 Channel capacity and codes

3.2.1 Discrete memoryless channels and capacity

In a *discrete memoryless channel* the input and output are sequences of symbols, and the current output depends only on the current input. Thus the channel is described by a pair of random variables, (X, Y), with values from finite alphabets $\mathcal{A}_x = \{x_1, x_2, \ldots, x_r\}$ and $\mathcal{A}_y = \{y_1, y_2, \ldots, y_s\}$. The conditional probability of y_j given x_i is $P(y_j|x_i) = q_{ij}$. We now introduce a concept of fundamental importance: the *capacity* of a discrete channel, $C(Y; X)$, is the maximum of $I(Y; X)$ with respect to $P(X)$. As discussed above, this definition is not surprising in itself, since, if one accepts the meaning of mutual information, the input symbols should be chosen in a way that maximizes this quantity.

In many cases of interest the maximum is easily determined. Thus the symmetry of the transition probabilities often suggests a symmetry in the input probabilities, and if necessary a maximization over a few parameters can be performed numerically. In the general case, analytical solutions are often difficult, as is demonstrated in a few problems.

Example 3.1 (The binary symmetric channel (BSC)). *We have already presented the BSC in Chapter 1. This channel models a situation in which random errors occur with probability p in binary data. The transition matrix is*

$$Q = \begin{bmatrix} 1-p & p \\ p & 1-p \end{bmatrix}.$$

For equally distributed inputs, the output has the same distribution, and the mutual information may be found from (1.5) as $1 - H(p)$ (in bits/symbol), where H indicates the binary entropy function. If one of the inputs is used with probability q, we find the mutual information as

$$I(X; Y) = H(Y) - H(p)$$

and the maximum is seen to occur when Y is equally distributed. This happens only for $q = 1/2$ unless $p = 1/2$. Thus the capacity is actually

$$C = 1 - H(p). \tag{3.1}$$

Example 3.2 (The binary erasure channel (BEC)). *If the channel erases symbols outputting a special character, $?$, but actual errors do not occur, the transition matrix becomes*

$$Q = \begin{bmatrix} 1-p & p & 0 \\ 0 & p & 1-p \end{bmatrix}.$$

The capacity of this channel, the binary erasure channel (BEC), is easily found from the last version of (1.6) as

$$C = 1 - p.$$

Thus the information is simply reduced by the fraction of the transmitted symbols that are erased.

Example 3.3 (The q-ary symmetric channel). *On this channel we can transmit one of q symbols, and in all cases the error probability is p. When an error occurs, the remaining $q - 1$ symbols each occur with probability $p/(q - 1)$. Again we use the input symbols* *with equal probability, and the output has the same distribution. The mutual information may be found from (1.6) as*

$$C = \log q - H(p) - p \log(q - 1),$$

where the entropy function can be interpreted as uncertainty about the position of the errors and the last term as the uncertainly about the error values.

3.2.2   Codes for discrete memoryless channels

It is clear that, for a memoryless channel, we can reach the maximal value of I by using independent inputs. However, it is not clear how these symbols can be used for communicating a message efficiently. If we want to transmit k input symbols reliably, and thus transfer $kH(X)$ bits of information, we must use the channel at least $n = kH(X)/C$ times in order to get enough information at the output.

An (n, k) *block code* is a set of vectors of length n, where k symbols are selected to represent the message, and the remaining symbols are functions of the information symbols. Thus, when a code is used, the transmitted symbols are no longer independent, and it would appear that a smaller amount of information might be transmitted. One way of understanding this situation is to convert the channel into a vector channel. Now that the input is one of the message vectors, the transition probabilities are found as the product of the symbols' probabilities since the channel is memoryless. We can now derive some information from (1.6). Assume that we want to get a mutual information close to $kH(X) = nC$. One version of the equation gives that

$$H(X_1, X_2, \ldots, X_n) = kH(X)$$

so $H(X_1, \ldots | Y_1, \ldots)$ should be zero. Thus an output vector should almost always indicate a unique transmitted message. The other version gives

$$nC = H(Y_1, Y_2, \ldots, Y_n) - nH(Y|X).$$

Thus the output distribution should be close to that of n independent symbols. The channel spreads out a transmitted vector on a set of received vectors, but the objective of the code design is to keep the overlap between these sets very small while filling up most of the space of received symbols.

The receiver must decode the received string to recover the information. The most common situation is one in which all messages are assumed to be equally likely, and we minimize the probability of an erroneous decision by decoding the vector y as the code vector x which maximizes $P[y|x]$. Such a decision is called a *maximum-likelihood* decision.

Unfortunately, although it can be proved that good codes exist, both the code construction and the decoding are extremely difficult. In the following section we consider codes for the BSC in more detail.

3.3   Linear codes

3.3.1   Linear codes and vector spaces

A binary linear block code of length n consists of k information bits and $n - k$ parity checks. We shall often assume that the information bits are transmitted unchanged, and the encoding is said to be *systematic*, but usually we do not need to discuss how the mapping from the information sequence to the corresponding codeword is chosen (although it is obviously important to specify it when the code is used). The $n - k$ parity bits are chosen in such a way that a certain subset of the transmitted bits contains an even number of 1s; we say that the set has even parity or that the word satisfies a parity check. This terminology is partly historic, but parity checks lead to the important notion of linear codes which covers most codes of practical importance.

A *linear* (n, k) *code* is a linear vector space of dimension k in the space of binary n vectors.

We shall use the symbols 0 and 1 to indicate the two values of the coordinates, and addition of these values is performed modulo 2 (multiplication by the constants 0 and 1 is trivial). We notice that this small system of numbers is a *field*, and that we can use most of the results from linear algebra over the rational numbers (or the reals). Where significant differences occur, we shall usually point them out.

A code may be specified by a *basis* of k linearly independent vectors. It is convenient to arrange these as rows in a k by n *generator matrix*, G. If the information is given in the format of a vector of length k, we can state the encoding rule as

$$y = uG.$$

It is often convenient to talk about the *rate* of a code, $R = k/n$.

The same code can be described by many different generator matrices, and we shall usually just select one that is convenient for our purpose. However, we notice that the generator matrix may also be interpreted as specifying a particular encoding of the information. Since G has rank k, we can reduce it by row operations in such a way that k columns form a k by k unit matrix, E. We shall often assume that this matrix can be chosen as the first k columns and write the generator matrix as

$$G = [EA].$$

This form of the generator matrix gives a systematic encoding of the information. We may now define a parity check as a binary vector, h, which satisfies

$$Gh = 0.$$

The parity-check vectors are again a linear vector space. The dimension is $n - k$, and we may express a basis for this space by collecting the rows to form an $n - k$ by n matrix, H, called a *parity-check matrix* for the code. From the systematic form of the generator matrix we may find H as

$$H = [A'E],$$

where A' denotes the transpose of A.

Example 3.4 ((8, 4) Code). *A short code is defined by the following generator matrix:*

$$G = \begin{bmatrix} 1 & 0 & 0 & 0 & 0 & 1 & 1 & 1 \\ 0 & 1 & 0 & 0 & 1 & 0 & 1 & 1 \\ 0 & 0 & 1 & 0 & 1 & 1 & 0 & 1 \\ 0 & 0 & 0 & 1 & 1 & 1 & 1 & 0 \end{bmatrix}.$$

By adding the four rows of G we get the all-1s vector. Thus the complement of a codeword is again a codeword. Since solving for H gives the complement of G, we find that the dual code is the same space as the code itself.

For a given received word (possibly containing errors), r, we define the *syndrome* by

$$s = Hr'. \tag{3.2}$$

We note that, if $r = y + e$, where y is a codeword and e is the error pattern, then

$$s = He.$$

The term syndrome refers to the fact that s reflects the error in the received word. The codeword itself does not contribute to the syndrome, and for an error-free codeword $s = 0$.

Since H also defines a linear space, we may interpret the parity checks as another code, the *dual code* to the original. We notice that, although the dimensions of the two dual spaces add to n as in the case of real vector spaces, a binary vector may be orthogonal to itself, and in fact the entire dual space may be part of the original space.

3.3.2   Decodable errors

An important indicator of the quality of an error-correcting code is the *minimum distance*, d, between two different codewords measured as the number of coordinates where the two words differ. For a linear code the sum of the two words is again a codeword, and we can find the minimum distance as the *minimum weight*, the number of nonzero coordinates, of a nonzero word. Often the terms *Hamming weights* and Hamming distances are used in honor of one of the founders of coding theory. The parameters of a code are written (n, k, d) if we want to include the minimum distance.

A code with minimum distance d can *correct any pattern of t or fewer errors*, where

$$t < d/2. \tag{3.3}$$

Suppose there were two different codewords, u and v, with distance less than $d/2$ from the received word. Then the weight of the codeword $u + v$ would be less than d, contradicting the statement that d is the minimum weight.

If we consider a codeword of weight d, we may select more than half of the 1s as errors, and the decoder would have to make a wrong decoding decision. Thus there are error patterns of any weight $> d/2$ that cause decoding errors. On the other hand, there may also be heavier error patterns that are correctly decoded.

Using the notion of syndromes from the previous section, we see that we can interpret a decoding as a mapping of syndromes to error patterns. Thus for each syndrome we should choose an error pattern with the smallest possible number of errors, and if there are several errors of equal weight we may choose one of these arbitrarily. Not only is such a syndrome decoder a useful concept, but also, if $n - k$ is not too large, decoding by table look-up may be a reasonable implementation.

On comparing the number of distinct syndromes and the number of correctable error patterns, we get the *Hamming bound*: the number of errors that can be corrected by a binary (n, k) code is upper bounded by the largest value of t such that

$$\sum_{j=0}^{t} \binom{n}{j} \leq 2^{n-k}. \tag{3.4}$$

Although it is usually not possible to find binary codes with minimum distances $2t + 1$ as suggested by this bound, it is often possible to correct most errors of weight t. In fact the analysis of long codes in Section 3.4 shows that this is the case for long randomly chosen codes.

For the total set of binary n-vectors we have the weight distribution

$$A(w) = \binom{n}{w}.$$

Many properties of particular codes are difficult to analyze, whereas it is possible to find averages over suitable sets of codes. For such a typical code (or randomly chosen code), we have

$$A(w) \approx 2^{k-n} \binom{n}{w}$$

for $w > 0$. Clearly $A(0) = 1$ for a linear code. For particular values of (n, k) and small w, the approximation is less than 1, and, since $A(w)$ is an integer, most codes do not have any words of such low weight. Thus there exist codes with minimum weight at least d, where d is the smallest integer such that

$$\binom{n}{d} \geq 2^{n-k}.$$

If a code is constructed by random selection of the codewords, it will obviously not often be linear. We can get a lower bound for linear codes by virtue of the following observation: if the minimum distance of a linear code is at least d, then all combinations of fewer than d columns in the parity-check matrix are linearly independent. If j columns sum to zero, there is a word, v, of weight j with nonzero coefficients in the corresponding positions such that $Hv = 0$. Consider constructing the parity-check matrix for a code with $n - k$ parity symbols by adding one column at a time (thus $n - k$ is fixed, but n and k increase). In each step we calculate the sum of all combinations of fewer than $d - 1$ columns. As long as there is a column different from these sums, we may use it to extend H. Clearly this is possible as long as

$$\sum_{j=0}^{d-2} \binom{n-1}{j} \leq 2^{n-k}. \tag{3.5}$$

Thus we have the *Gilbert–Varshamov lower bound* on minimum distance: there exists a binary linear (n, k) code with minimum distance d as long as (3.5) is satisfied.

We shall often need the approximation

$$\log \binom{n}{w} \approx n H(w/n), \tag{3.6}$$

where H is the binary entropy function. By using this approximation in the expression for the weight distribution and taking logarithms we get a bound for long codes, namely the *Gilbert bound* (asymptotic version): there exist long binary (n, k, d) codes when d/n and the rate $R = k/n$ satisfy

$$H(d/n) < 1 - R. \tag{3.7}$$

It is not known whether there exist long binary codes with fixed rate $0 < R = k/n < 1$ and larger minimum distances. Also there is no known construction (with complexity polynomial in n) for constructing long codes that reach this bound. It may be noted that the upper bound gives a distance that is exactly twice as large, namely the *Hamming bound* (asymptotic version): the relative minimum distance d/n of a long binary (n, k, d) code is upper bounded by $d/n < 2t/n$, where

$$H(t/n) < 1 - R, \tag{3.8}$$

since we get the same expression with d replaced by t, and $d > 2t$. Since the expected number of errors on a long block is np, the Hamming bound indicates that the errors cannot be corrected if the rate of the code exceeds the capacity of the BSC.

3.3.3 Hamming codes and their duals

If a single error occurs, the syndrome equals the corresponding column of H. Thus we may define a single-error-correcting code in the following way: *a Hamming code* is defined by taking H as an m by $2^m - 1$ matrix in which all columns are different and nonzero.

It follows immediately that single errors are corrected since the syndromes are distinct, and thus we may also conclude that d is at least 3. It is easily verified that $d = 3$, since we can choose three positions such that the columns of H sum to 0.

An extended Hamming code is defined by adding a zero column to the parity-check matrix of a Hamming code and then adding a row of all 1s.

The minimum distance of the extended Hamming code is 4. A single error in the last position gives a zero syndrome in the original m places, but the row of 1s always gives a syndrome bit of 1 for a single error. On the other hand, the extra parity check makes all codewords have even weight. Thus the parameters are $(2^m, 2^m - m - 1, 4)$.

An orthogonal code is the dual of a Hamming code. An orthogonal code consists of the all-0s vector and n vectors of weight $(n + 1)/2$. A *biorthogonal code* is the dual of an extended Hamming code. This code consists of the all-0s vector, the all-1s vector, and $2n - 2$ vectors of weight $n/2$.

Table 3.1. Generators of some binary cyclic codes (in hexadecimal representation)

(n, k, d)	$g(X)$
$(15, 11, 3)$	13
$(21, 12, 5)$	337
$(23, 12, 7)$	AE3
$(31, 26, 3)$	25
$(31, 21, 5)$	769
$(31, 16, 7)$	8FAF

3.3.4 Cyclic codes and generator polynomials

So far we have not specified any particular order of the code symbols, and equivalent versions of the codes can clearly be obtained by reordering the columns of the G and H matrices. In many codes of interest it is possible to choose a generator matrix of the form

$$\begin{bmatrix} g_{n-k} & g_{n-k-1} & \cdots & 0 & 0 \\ 0 & g_{n-k} & \cdots & 0 & 0 \\ \vdots & \vdots & \ddots & \vdots & \\ 0 & 0 & \cdots & g_1 & g_0 \end{bmatrix}.$$

Thus, if the positions of the code are associated with powers of an indeterminate, $x^{n-1}, x^{n-2}, \ldots, x, 1$, the last row represents the *generator polynomial*, $g(x)$, the other rows are obtained as $x^j g(x)$, and any codeword is a multiple of $g(x)$ when interpreted as a polynomial. This approach allows a much more compact notation, but it also leads to a more efficient implementation.

An information sequence, $u(x)$, can clearly be encoded as $u(x)g(x)$, but it is often desirable to use the less obvious systematic encoding

$$u(x)x^{n-k} - (u(x)x^{n-k} \bmod g(x)).$$

Thus the information sequence is placed on the left, and the remainder after division by $g(x)$ is appended as the parity part. The syndrome can always be found by dividing the received sequence by $g(x)$.

A *cyclic code* is obtained when $g(x)$ divides $x^n - 1$. In such a code any cyclic shift of a codeword is again a codeword, since we can add a multiple of $x^n - 1$ to move any coefficients of $x^j g(x)$ of degree greater than $n - 1$ to the low-order positions.

For a cyclic code the dual is also cyclic.

A short list of generator polynomials and parameters of some good short cyclic codes is presented in Table 3.1. These codes may be extended by an overall parity check to get codes of even distance.

Example 3.5 (Hamming code). *We may list the 15 nonzero columns of length 4 by taking binary representations of the integers from 1 to 15, but a cyclic code is obtained*

by using the contents of a shift register with linear feed-back defined by the polynomial
$x^4 + x + 1$,

$$
H = \begin{bmatrix}
1 & 0 & 0 & 0 & 1 & 0 & 0 & 1 & 1 & 0 & 1 & 0 & 1 & 1 & 1 \\
0 & 1 & 0 & 0 & 1 & 1 & 0 & 1 & 0 & 1 & 1 & 1 & 1 & 0 & 0 \\
0 & 0 & 1 & 0 & 0 & 1 & 1 & 0 & 1 & 0 & 1 & 1 & 1 & 1 & 0 \\
0 & 0 & 0 & 1 & 0 & 0 & 1 & 1 & 0 & 1 & 0 & 1 & 1 & 1 & 1
\end{bmatrix}.
$$

This gives a parity-check matrix for a (15, 11, 3) *Hamming code. Since the matrix contains a 4 by 4 unit matrix, we can obtain a generator matrix as discussed above. The same matrix is a generator matrix for a* (15, 4, 8) *orthogonal code. By adding a zero column and a length-16 all-1s vector to the matrix, we get a generator matrix for the* (16, 5, 8) *biorthogonal code. The dual* (16, 11, 4) *extended Hamming code may be found by adding a column to the generator matrix of the* (15, 11, 3) *code and making the weights of all rows even.*

As demonstrated in the example above, the Hamming codes and the orthogonal codes are cyclic. The extended Hamming codes are not cyclic, but a similar code of length $2^m - 1$ can be obtained by multiplying the generator polynomial of the Hamming code by $x - 1$. Such codes are often used as parity checks on long files, called *cyclic redundancy checks (CRCs)*. Such codes are not used for error correction, but it is verified that the remainder (syndrome) is zero as a way of assuring the integrity of the data.

3.3.5   Orthogonal sequences and arrays for synchronization

When an orthogonal code is written in cyclic form, the n nonzero codewords are exactly all cyclic shifts of a single sequence (called a *maximum-length sequence* or *m*-sequence). The name refers to the fact that this is the longest sequence that can be generated by a linear shift register of length m with linear feed-back. A sum of two codewords is again a codeword, and consequently we see that they differ in exactly $(n + 1)/2$ places. This property makes such sequences useful for obtaining synchronization of transmitted frames, which is again an important necessary condition for proper application of decoders of error-correcting codes and source codes. Other applications include various forms of measurements of impulse responses and delays (radar, ultra-sound).

In a two-dimensional (2-D) page it is similarly necessary to have markers to ensure proper alignment of the medium before it can be read (by a machine). If a document is scanned before such an alignment can be obtained, a suitable correction of the data must be computed.

The markers often have a simple structure in practical systems, but an interesting pattern can be derived from orthogonal codes. Let the length of the code be a product of two mutually prime numbers, $n = uv$, and write the sequence into a u by v array along the diagonals. Thus bit i in the sequence is written into position $(i \bmod v, i \bmod u)$ in the array. Any 2-D cyclic shift of the array can now be obtained as some one-dimensional (1-D) cyclic shift of the orthogonal sequence. Thus the shifted array differs from the original in $(n + 1)/2$ places as in the 1-D case.

Orthogonal 2-D arrays also have other applications, including imaging based on sensor arrays placed behind a rectangular mask where a 0 in the binary array indicates a transparent cell and 1 an opaque cell.

3.3.6 Product codes

A 2-D code can be obtained by considering an array of data and applying error-correcting codes to rows and columns. The use of such codes is sometimes motivated by the nature of the data and the error mechanisms that affect the transmitted words, but they may also be used to get long codes that can be decoded by several applications of simple decoding methods.

Given two linear codes with parameters (n_1, k_1) and (n_2, k_2), the *product code* consists of binary arrays such that all rows are codewords in the first code and all columns are codewords in the second code.

The parameters of the product code are

$$(n, k) = (n_1 n_2, k_1 k_2, d_1 d_2).   \tag{3.9}$$

The length is obvious. The number of parity checks is at most $n_1(n_2 - k_2) + n_2(n_1 - k_1)$, but, when the same code is used in all rows and columns, we can choose the first k_1 rows as codewords in the first code, and encode the columns by taking the right linear combinations of these rows. In any nonzero array there are at least d_1 nonzero columns, each with at least d_2 symbols. Thus the minimum distance is at least the product of the two distances, but, if the codes are the same in all rows and in all columns, the same minimum-weight words can be repeated to give an array with exactly that weight. As suggested by this discussion, it is possible to get codes of lower rate but larger distance by permuting the positions in rows and columns.

The minimum distances of product codes are disappointing, but it is possible to correct more than $d_1 d_2/2$ errors in many situations. We can describe a large class of decodable errors in the following way. Let the row and columns codes have the same minimum distance, d. If the number of errors in any intersection of $m \geq d$ rows and columns is less than $md/2$, the error pattern can be corrected by the product code. To see that no two such error patterns have the same syndrome, we add them and disregard any rows and columns where the two patterns are identical. The remaining m' nonzero rows and columns have weight less than $m'd$, and thus cannot be a codeword in the product code. Consequently the two syndromes cannot add to zero.

When the error pattern satisfies this condition, there is at least one row or column with a nonzero error pattern of weight less than $d/2$. Thus the errors can be corrected by repeated application of an algorithm that corrects errors of weight less than $d/2$ errors in the component code. However, since some of these corrections may be decoding errors, the decoder must perform a tree search until the unique error pattern satisfying the weight condition and all row and column syndromes has been found.

A more practical approach to decoding is to repeat the decoding of rows and columns several times. The performance of this method is discussed in Chapter 8.

3.4   Error probability and error exponents

This section presents some results on error probabilities for codes on the binary symmetric channel (BSC). The presentation is based on the classical technique of using randomly chosen codes for discrete channels. However, we have simplified the derivation by considering a specific channel, namely the BSC. One of the aims of this section is to show that it is possible to communicate with arbitrarily small error probability at any rate less than capacity. Although the construction of linear codes provides some of the tools for achieving such a result, it is not yet possible to construct optimal long error-correcting codes, and it is very difficult to analyze the performance of specific codes.

Here we follow Shannon's approach of considering the average performance of a randomly chosen code. This argument in itself, and the emphasis on asymptotic bounds and exponential terms in the approximations, may require more mathematical maturity than other sections of the chapter. We provide an example at the end of the section to demonstrate that the results agree with direct evaluations for codes of moderate lengths.

3.4.1   The union bound for block codes

We consider a code consisting of 2^k vectors of length n. The rate of the code is $R = k/n$. The transition probability on the BSC is p (see Chapter 1), and the number of codewords at distance w from the transmitted word is $A(w)$. We assume that a received vector is decoded as the closest codeword *maximum-likelihood decoding*, but we shall not discuss decision between several equally likely codewords.

An error occurs if the received word is closer to some word different from the one that was transmitted. For each word at distance $w > 0$, the probability that the received word is closer to this word is

$$\sum_{j > w/2} \binom{w}{j} p^j (1 - p)^{w-j}$$

since there must be errors in more than $w/2$ of the positions where the two words differ. We now get an upper bound on the probability of the union of these events by taking the sum of their probabilities. Thus the error probability is upper bounded by

$$P(e) < \sum_{w > 0} \sum_{j > w/2} A(w) \binom{w}{j} p^j (1 - p)^{w-j}.$$

Since most errors occur for j close to $w/2$, we may use the approximation

$$P(e) < \sum_{w > 0} A(w) 2^w (p - p^2)^{w/2}.$$

On introducing the function $Z = \sqrt{4p(1 - p)}$, which depends only on the channel, we get

$$P(e) < \sum_{w > 0} A(w) Z^w.$$

In order to study how the error probability depends on n, we shall make some further approximations. The distance distribution is often not known, but it may be approximated by the average distribution, which is obtained by scaling the binomial distribution to the right number of codewords:

$$A(w) = \binom{n}{w} 2^{k-n}.$$

We use the approximation to the binomial coefficient (3.6), and, since the summation is over N terms, we get the right exponent by taking the largest term:

$$\log P(e) \leq \max[k - n + nH(w/n) + w \log Z].$$

The maximum is obtained for $w/n = Z/(1 + Z)$. Thus there exist block codes of length n and rate R such that the error probability on a BSC is upper bounded by

$$\log P(e) \leq -n(R_0 - R), \tag{3.10}$$

where

$$R_0 = 1 - \log(1 + Z).$$

For a fixed rate, the error probability decreases exponentially with the length of the code, and the exponent is bounded by a straight line with slope -1. Often the term error exponent is used for $E(R) = R_0 - R$. This result is both fairly simple and quite powerful. It is also the first step toward the *channel coding theorem*.

The union bound gives a useful result only for $R < R_0$. We shall see in the next section that it is possible to obtain a positive error exponent for any rate less than the channel capacity, $C = 1 - H(p)$, by use of a different technique.

Where this error exponent, $E(R) = R_0 - R$, applies, most of the errors are associated with codewords of weight w/n, which is a function of the channel, but not of the rate of the code. In addition to the errors that hit the w/n nonzero bits of the codeword, an expected $(1 - w/n)p$ errors occur in other positions.

3.4.2 A bound for rates close to capacity

For rates close to the channel capacity, which in some sense is the most interesting case, the error probability is over-estimated in (3.10). As discussed previously, the largest contribution to the error probability comes from patterns of weight $j = w/2 + (n - w)p$, where $w/2$ errors occur in positions where a weight w word has 1s, and the $(n - w)p$ errors occur in some other positions. The bound can be expressed in more detail as

$$\sum_{w>0} A(w) \left[\sum_{j>w/2} \binom{w}{j} p^j (1-p)^{w-j} \sum_{i=0}^{n-w} \binom{n-w}{i} p^i (1-p)^{n-w-i} \right].$$

If these terms are collected by the total number of errors, u, we get

$$\sum_{w>0} A(w) \sum_{u>w/2} \sum_{j>w/2} \binom{w}{j}\binom{n-w}{u-j} p^u(1-p)^{n-u}.$$

For large u the same error patterns are counted several times, and we get a better bound by assuming that all errors of weight $u > U$ cause decoding error,

$$\sum_{u>U} \binom{n}{u} p^u(1-p)^{n-u}.$$

Using the same asymptotic approximations as before with $x = u/n$, we get the exponent

$$-n(-H(x) - x\log p - (1-x)\log(1-p)) \tag{3.11}$$

and the exponent is positive as long as $x > p$. The best bound is obtained with the smallest value, U/n, for which this expression applies. This is the value of the Hamming bound

$$H(x) = 1 - R$$

since a larger number of errors would certainly cause decoding errors in almost all cases.

By combining $w/n = Z/(1+Z)$ and $j = w/2 + (n-w)p$ we find that the exponent in (3.10) is actually the tangent to the one in (3.11), and the two expressions coincide for the "critical" fraction of errors

$$u/n = \frac{\sqrt{p}}{\sqrt{p} + \sqrt{1-p}}.$$

For higher rates, (3.11) with the given value of x is a lower bound on the error exponent (upper bound on the error probability), and we have a proof of the coding theorem for the BSC. Actually no code could have a better error exponent, since the expression would apply to a code that could correct all error patterns within the Hamming bound. The derivation shows that for an average code the words at distance less than the Hamming bound do not contribute to the error exponent. Thus for rates in this range we have the correct exponent, even though the best minimum distance is not known.

As R gets close to C, the exponent is small, and it is difficult to get a low probability of decoding error.

Example 3.6 (The error exponent for the BSC).   *Let the error probability on the channel be $p = 0.1$. The channel capacity for this channel is $C = 1 - H(0.1) = 0.531$. We then have $Z = 0.6$, $R_0 = 1 - \log 1.6 = 0.322$. On the straight-line bound we have $w/n = Z/(Z+1) = 0.375$. Thus, in the union bound, the error probability is dominated by codewords at this distance for all rates to which the bound applies. At low rates we know from the Gilbert bound that some codes do not have words of such low weight, and a better bound could be found. However, this happens for code rate $R = 1 - H(0.375) = 0.0455$, and we omit the details. The critical fraction of errors is $1/4$, and from this we get that* *the critical rate (at which the high-rate bound starts) is* $1 - H(0.25) = 0.189$. *Between this rate and the capacity the error exponent is*

$$R - 1 - x \log p - (1 - x)\log(1 - p)$$

for

$$H(x) = 1 - R.$$

For $x = 0.2$ *we find* $R = 0.278$ *and* $E = 0.064$, *for* $x = 0.15$, $R = 0.390$, *and* $E = 0.018$.

3.4.3 Performance of specific codes on the BSC

The bounds in the previous sections apply to long average codes, but the results suggest what performance we can expect from codes of moderate length. While the asymptotic bounds use a fixed channel and codes of increasing length, it is more common in applications to fix the code and let the error probability on the channel vary.

For given parameters (n, k), the largest number of errors that can be corrected is given by (3.4). Thus for $p = t/n$ the code rate equals the capacity of the channel, and we can expect a useful performance only for lower values of p. However, the high-rate error exponent indicates that for p close to this value we can estimate the probability of decoding error by the probability that a received word contains more than t errors. Thus for such channels the performance does not depend on the minimum distance of the code. However, when p is small and there are very few decoding errors, the error probability is dominated by the union bound, and the performance depends on the weight distribution of the code. Eventually only minimum-weight codewords will contribute to the error probability.

The bounds assume that maximum-likelihood decoding is used. This is usually not possible for long codes. However, it is useful to compare the actual performance of a real decoder (simulated or calculated) with the optimal performance.

Example 3.7 (Error probabilities). *For a moderate code length of* $n = 256$ *and* $R = 1/2$, *the Hamming bound indicates that at most 29 errors can be corrected, which is very close to the asymptotic expression. Similarly, we get from the Gilbert bound that there are codes with minimum distance at least 30, but actually better codes exist. Assume that 16 errors are corrected, but that there are words of weight 33. From the binomial approximation we expect that the code contains about 1000 such words. Even for* $p = 1/16$, *the probability of error due to these codewords is only* 10^{-14}. *Thus, even though only half of the received codewords can be decoded if we decode 16 errors, there are few decoding errors. If we are able to decode more errors, we would find a much larger number of decoding errors. Since* $Z = 0.484$, *the analysis in this section indicates that most errors are related to codewords of weight about 85, and thus error patterns of weight about 43. A direct calculation verifies that this case gives an error probability of about 0.05. However, for* *this rate we get a better approximation by assuming that all error patterns of weight 30 cause decoding error. In this way the error probability is estimated as 2×10^{-4}.*

3.5   Other applications of linear codes in information theory

Linear codes were developed to correct errors on binary channels, and as such they serve to demonstrate the relevance of the concepts of mutual information and capacities of discrete channels. However, several other problems in information theory require codes with similar properties. In this section we briefly mention some applications of linear binary codes in other areas of information theory.

3.5.1   Rate-distortion functions

In the first chapters we considered the encoding of a source such that it is possible to restore the original sequence without error or at least with a very small probability of error. In rate-distortion theory we consider the more difficult question of how much information is needed in order to restore the source sequence with a prescribed error.

An analysis of this problem requires a description of the source as well as a measure of the distortion between the original and the reproduced source. In general the reproduced sequence could use a different alphabet (as in quantization of a continuous source), and we have to associate a cost with replacing a particular source symbol by a certain reproducing symbol.

For an original source variable, X, a reproduced variable, Y, and a certain average distortion $D(X, Y)$, we want to determine the minimum of $I(X; Y)$. This function of D is called the *rate-distortion function*, $R(D)$. It generalizes the entropy function by specifying the minimum rate at which the source symbols on average can be reproduced without exceeding the specified distortion, D. For $D = 0$ the reproduction is lossless and $R(D = 0) = H(X)$.

Here we consider only the simple case in which the source is memoryless and binary and the reproducing alphabet is the same. The distortion will be measured simply as the probability that a reproduced symbol differs from the original.

Let the probability distribution for the source be $P[X = 1] = p \leq 1/2$, without loss of generality. We will show that the rate-distortion function is

$$R(D) = H(p) - H(D) \tag{3.12}$$

for $D < p$ and 0 otherwise. We first give a lower bound on I,

$$I(X; Y) = H(X) - H(X|Y) \geq H(p) - H(D),$$

since the conditional entropy, $H(X|Y)$, cannot exceed the entropy of a memoryless sequence of errors with $P[1] = D$.

It is not easy to state how to specify and code Y for a given X. However, we can define a channel with input Y and output X such that the bound is achieved, thus minimizing I.

In this channel the roles of input and output are reversed, and we consider Y as the input and X as the output of a BSC with transition probability D, and we choose the input distribution to be $(q, 1 - q)$, where

$$p = D(1 - q) + (1 - D)q,$$

$$q = \frac{p - D}{1 - 2D}.$$

The distributions of X and Y are as specified, and the mutual information meets the lower bound. This is a rather complicated argument considering the fairly simple problem, and even so it is not very satisfying, since it does not indicate how the source should be encoded.

If we take the easier case $p = 1/2$, a linear code does provide the answer: we can take a source sequence of length n and decode it using a linear code of rate R. The result is an information sequence of length K, which we store. The source can then be reproduced by encoding the information sequence. From the capacity of the binary channel, it follows that this process is possible for $R > 1 - H(q)$, which is what we get from the rate-distortion function.

3.5.2 Encoding of correlated sources

If we have two source sequences (perhaps obtained as noisy observations of a common target), X and Y, their total entropy can be expressed by the chain rule as

$$H(X, Y) = H(X) + H(Y|X).$$

We could encode this information using a number of bits close to the entropy by applying a suitable-source-coding procedure to the observed pair of sequences.

The observation in Slepian–Wolf coding is that an encoding with the same number of symbols is possible even when X and Y have to be processed separately, for instance before they are transmitted to a common receiver.

Again we treat only the simple binary memoryless case. Both X and Y are binary sequences of independent symbols, but they differ from each other only with probability $p < 1/2$. The X sequence is stored (or transmitted) as observed. The other sequence is decoded using a linear code of rate $R > 1 - H(Y|X)$, and the syndrome is stored (transmitted). The original sequence can be recreated with high probability by finding the minimum-weight pattern corresponding to the syndrome (the error pattern in decoding). This pattern is added to X, the result is decoded using the same code, and the error pattern is again added to the result (note that this is not necessarily the same as just decoding the X sequence and adding the error sequence). Since the same "error" sequence was added to X and Y, it did not change their mutual distance, and the decoding of X changes only a fraction of positions, q, such that $H(q) < H(Y|X)$. Thus, if the decoding was correct, we have recreated the correct Y sequence.

3.6   Notes

The are several books devoted to the theory of error-correcting codes. Some of these approach the topic primarily from a mathematical point of view. References [1] and [2] were written for an audience with a background in electrical engineering or computer science.

Exercises

3.1 Find the capacities of binary symmetric channels with error probabilities $p = 0.001$, $p = 0.01$, and $p = 0.1$.

3.2 Find the capacity of a channel with errors and erasures given the following transition matrix (and assuming that the two input symbols are used with probability $1/2$):

$$Q = \begin{bmatrix} 2/3 & 1/4 & 1/12 \\ 1/12 & 1/4 & 2/3 \end{bmatrix}.$$

3.3 Find the capacity of the "Z-channel" with transition matrix (using analytical or numerical maximization over the input distribution)

$$Q = \begin{bmatrix} 4/5 & 1/5 \\ 0 & 1 \end{bmatrix}.$$

How much is the mutual information reduced if equally likely input symbols are used?

3.4 Find the capacity of a channel with 16 input and output symbols assuming that the correct symbol is received with probability 0.9, while for some numbering of the symbols, numbers $i - 1$ and $i + 1$ modulo 16 are received with probability 0.05.

3.5 Find the minimum distance of the $(8, 4)$ code in Example 3.4. How many errors can the code correct?

3.6 Compare the minimum distances of the codes in Table 3.1 to the Hamming and Gilbert bounds.

3.7 In Example 3.5 we mentioned the $(16, 11, 4)$ extended Hamming code.
(a) What are the parameters of a product of two such codes?
(b) How many errors can be corrected?

3.8 In Example 3.5 an m-sequence of length 15 was presented as a row of H. Convert this sequence into a 5 by 3 array that is orthogonal to cyclic shifts of rows and columns.

3.9 Assume that we want to use a binary $(64, 32)$ code on a BSC with $p = 0.01$.
(a) What is the capacity of the channel?
(b) Find R_0 and evaluate the union bound on error probability using the asymptotic expression.
(c) Find an approximate weight distribution for an average code with the given parameters. Compare the result with the Gilbert bound on minimum distance and the Hamming bound on the number of decodable errors.

    (d) Assuming the approximate weight distribution, find the weight of the error patterns that contribute most to the error probability. Compare the result with the value obtained before using the error exponent.

3.10 Consider a long sequence of random binary data.

    (a) How much can it be compressed if we allow 3% errors in the restored sequence?

    (b) If a Hamming code of length 31 is used for the compression (storing only the information bits), what is the resulting error rate? How many bits are stored?

3.11 Two sensors record a set of correlated data. Assume that the two bit strings, X and Y, differ in 5% of places, randomly distributed.

    (a) Find $H(Y|X)$. X is transmitted directly to the common receiver, but Y is decoded using a binary $(31, 21, 5)$ code.

    (b) How many errors can the code correct?

    (c) Explain how Y can be reconstructed from X and the syndrome if the number of disagreements is at most the number of errors corrected by the code.

    (d) What happens if the difference between the bit streams exceeds this value?

References

[1] S. Lin and D. J. Costello, *Error Control Coding, Fundamentals and Applications*, 2nd edn. (Upper Saddle River, NJ: Prentice-Hall, 2004).

[2] J. Justesen and T. Hoholdt, *A Course in Error-Correcting Codes* (Zurich: European Mathematical Society, 2000).

4 Reed–Solomon codes and their decoding

4.1 Introduction

Reed–Solomon codes are error-correcting codes defined over large alphabets. They were among the early constructions of good codes (1959), and are now one of the most important classes of error-correcting codes for many applications. At the same time these codes constitute a natural starting point for studying algebraic coding theory, i.e. methods of correcting errors by solving systems of equations.

4.2 Finite fields

To describe the codes and the decoding methods, the symbol alphabet is given a structure that allows computations similar to those used for rational numbers. The structure is that of a field. In a *field* there are two compositions, addition and multiplication, satisfying the usual associative and distributive rules. The compositions have neutral elements 0 and 1, every element has an additive inverse (a negative), and nonzero elements have a multiplicative inverse.

Well-known examples of fields include the rational, the real, and the complex numbers. The integers are not a field because only ± 1 have multiplicative inverses. However, there are also fields with a finite number of elements, and we actually used the binary field in the previous chapter to construct binary codes. A *finite field* with q elements is referred to as $F(q)$. Having the alphabet given this structure allows us to use concepts of matrices, vector spaces, and polynomials, which concepts are essential to the construction of codes and decoding algorithms.

The simplest examples of finite fields are sets of integers, $[0, 1, 2, \ldots, p - 1]$, with addition and multiplication modulo p. The only property that is not obvious is the existence of multiplicative inverses. Let x be a nonzero element and consider the products $0x, 1x, 2x, \ldots, (p - 1)x$. If two of these were identical modulo p, p would have to divide $(ix - jx)$, but that is impossible since both x and $i - j$ are smaller than p. Thus the products are distinct modulo p, and one of them has to equal 1.

Example 4.1 (The field $F(11)$).   *The integers from 0 to 10 can be viewed as elements of the field $F(11)$. Taking the results of addition and multiplication modulo 11, we get $7 + 8 = 4$, and $4 \times 5 = 9$. The element 6 has inverse 2, since $2 \times 6 = 1$, thus we can divide $5/6 = 5 \times 2 = 10$. This field is sometimes used for calculating parity checks on account numbers and other decimal strings. Combinations that give 10 as the result are not used.*

Most of the examples in this chapter use such prime fields. However, most applications involve encoding of binary data, and fields with 2^m elements are preferred. In particular, many standards use the field $F(2^8)$. We construct these fields in a later section.

Since it is not the purpose of this text to develop the mathematical theory of finite fields, we will comment on similarities and differences between results for the rationals and the finite fields in the relevant sections. In the previous chapter we defined codes by generator and parity-check matrices over $F(2)$. This approach is readily generalized to linear (n, k) codes over $F(q)$ by using vectors and matrices with elements from that field.

Example 4.2 (Singe-error-correcting codes in $F(q)$).   *A code similar to binary Hamming codes can be defined by the parity-check matrix, H. We use the field $F(11)$ in this example:*

$$H = \begin{bmatrix} 1 & 1 & 1 & 1 & 1 & 1 & 1 & 1 & 1 & 1 \\ 1 & 2 & 3 & 4 & 5 & 6 & 7 & 8 & 9 & 10 \end{bmatrix}.$$

For a received vector, r, the syndrome is calculated as $H \times r'$ (mod 11). The first element of the syndrome gives the error value; the second is the product of the value and the error location.

4.3   Definition of Reed–Solomon codes

Polynomials over $F(q)$ are expressions of the form $f_m z^m + f_{m-1} z^{m-1} + \cdots + f_1 z + f_0$, where the coefficients f_j are elements from $F(q)$, z is a variable, and the exponents are integers (not elements of $F(q)$). Thus the exponent is interpreted as a multiplication of z by itself m times. Polynomials in several variables are written similarly, and we can use well-known procedures for multiplying and dividing polynomials. Reed–Solomon codes can be defined as evaluations of certain polynomials:

Let $x_1, \ldots, x_n$ be different elements of a finite field $F(q)$. For $k \leq n$ consider the set of polynomials with coefficients in $F(q)$ and of degree less than k. A *Reed–Solomon* (RS) code consists of the codewords

$$(u(x_1), u(x_2), \ldots, u(x_n)).$$

It is clear that the length of the code is $n \leq q$. The code is linear since, if $c_1 = (f_1(x_1), \ldots, f_1(x_n))$ and $c_2 = (f_2(x_1), \ldots, f_2(x_n))$, then $ac_1 + bc_2 = (g(x_1), \ldots, g(x_n))$, where $a, b \in F(q)$ and $g(x) = af_1(x) + bf_2(x)$.

The polynomials form a vector space over $F(q)$ of dimension k since there are k coefficients. Two distinct polynomials cannot generate the same codeword since the difference would be a polynomial of degree $< k$ and it cannot have n zeros, so the *dimension* of the code is k. A codeword has weight at least $n - k + 1$ since a polynomial of degree $< k$ can have at most $k - 1$ zeros. Thus the *minimum distance* of an (n, k) RS code is $d = n - k + 1$.

4.4   Decoding by interpolation

We write the received vector as $r = c + e$, where the error vector e has at most $t = (n - k)/2$ nonzero entries. The idea is to determine a polynomial in two variables, $Q = Q_0(x) + yQ_1(x)$, with coefficients in $F(q)$ such that

$$Q(x_i, r_i) = 0, \quad i = 1, 2, \ldots, n.$$

The polynomial Q is called an interpolating polynomial since it agrees with all points of r. We shall constrain the degree of Q in such a way that it may have $n + 1$ coefficients. By treating these coefficients as variables, the n received points give us n homogeneous linear equations, and there is always a nonzero solution. The degrees of Q_0 and Q_1 are further chosen to satisfy

$$\deg(Q_0) \leq n - t - 1 = l_0, \qquad \deg(Q_1) \leq n - k - t = l_1$$

and, since $2t = n - k$, the sum of the degrees is at most $n - 1$.

We can write out the system of linear equations used in finding Q as

$$\begin{bmatrix} 1 & x_1 & x_1^2 & \cdots & x_1^{l_0} & r_1 & r_1 x_1 & \cdots & r_1 x_1^{l_1} \\ 1 & x_2 & x_2^2 & \cdots & x_2^{l_0} & r_2 & r_2 x_2 & \cdots & r_2 x_2^{l_1} \\ \vdots & \vdots & \vdots & \ddots & \vdots & \vdots & \vdots & \ddots & \vdots \\ 1 & x_n & x_n^2 & \cdots & x_n^{l_0} & r_n & r_n x_n & \cdots & r_n x_n^{l_1} \end{bmatrix} \begin{bmatrix} Q_{0,0} \\ Q_{0,1} \\ Q_{0,2} \\ \vdots \\ Q_{0,l_0} \\ Q_{1,0} \\ Q_{1,1} \\ \vdots \\ Q_{1,l_1} \end{bmatrix} = \begin{bmatrix} 0 \\ 0 \\ 0 \\ \vdots \\ 0 \\ 0 \\ 0 \\ \vdots \\ 0 \end{bmatrix}. \qquad (4.1)$$

If the received vector contains at most t errors, r_i differs from $c_i = g(x_i)$ in at most t places. It follows that $Q(x_i, u(x_i)) = 0$ for at least $n - t$ values of x_i. But, since

$\deg(Q(x, u(x))) \le n - t - 1$, it must be 0, thus $Q_0(x) + u(x)Q_1(x) = 0$. We can write the information polynomial as

$$u(x) = -Q_0(x)/Q_1(x). \tag{4.2}$$

A different approach to decoding follows from writing Q as $Q_1(y - u(x))$, which shows that Q_1 has zeros for those values of x that are error positions. We refer to this polynomial as an *error-locator polynomial*. In the following section we describe decoding methods that work with smaller systems of equations and determine this polynomial without using Q_0.

Example 4.3 (RS code in $F(11)$). *We encode a $(10, 5)$ code over $F(11)$ by evaluating an information polynomial of degree 4 for the ten nonzero values of z (it could also be evaluated for 0, but most often this point is not used). For $u(z) = 3 + 2x + x^4$ we find $c = (u(1), u(2), \ldots, u(10)) = (6, 1, 2, 3, 0, 2, 9, 1, 4, 2)$.*

Example 4.4 (Decoding by interpolation). *In this example we give a simple MATLAB script for decoding $t = 2$ errors in the $(10, 5)$ RS code over $F(11)$. If fewer than t errors occur, the coefficient matrix is singular, and a smaller value of t should be used. Clearly, if all syndromes are zero, no decoding is needed. In this script we use integer arithmetic to simplify the calculations.*

```
%Decoding by interpolation
p=11;                   %The symbol field is integers modulo p
n=p-1;                  %length of code
t=2;                    %number of errors corrected
ee=[6 1 2 3 0 2 9 1 4 2];          %received word
rr=[7 3 2 3 0 2 9 1 4 2];          %received word

%Set up homogeneous equations as (4.1)

mq(:,1)=ones(p-1,1);   %first column of coefficient matrix
mq(:,2)=[1:p-1]';      %second column of coefficient matrix
for ii=3:n-t,
    mq(:,ii)=mod(mq(:,ii-1).*mq(:,2),p);    %next columns
end
mq(:,n-t+1)=rr';                   %column of received values
for ii=1:t,
    mq(:,n-t+ii+1)=mod(mq(:,n-t+ii).*mq(:,2),p);%next columns
end                                %Coefficient matrix as (4.1)
```

```
%Find the solution by integer arithmetic mod p

dd=det(mq(:,1:n));                        %determinant
mqi=mod(round(dd*inv(mq(:,1:n))),p);
                    %The scaled inverse has integer entries
qq=mod(mqi*mq(:,n+1),p);                  %solve linear system
qq=[qq',mod(-dd,p)];              %Find the leading coefficient
div=mod(-dd*[1:p-1],p);                   %by inverting dd
fact=find(div==1);                        %product dd*fact = 1
qq=mod(qq*fact,p);                        %scale with inverse
q0=qq(n-t:-1:1);                          %Q0 polynomial
q1=qq(n+1:-1:n-t+1);                      %Q1 polynomial
inn=mod(-deconv(q0,q1),p)                 %use deconv to divide
```

4.5   Decoding from syndromes

Most decoders for RS codes are based on algorithms that require less computation than
the method described in the previous chapter. However, to describe these methods we
need an important concept from the theory of finite fields:

A *primitive element*, α, in the field $F(q)$ has the property that all nonzero elements
can be expressed as powers of α.

We may list the nonzero elements as consecutive powers,

$$\alpha^0 = 1, \alpha, \alpha^2, \ldots, \alpha^{q-2}$$

and $\alpha^{q-1} = 1$. Two elements can be multiplied by adding their exponents modulo $q - 1$.

Using this representation of the field elements, we introduce an important matrix:

$$F = (\alpha^{ij}), \quad i, j = 0, 1, \ldots, q - 2. \tag{4.3}$$

We refer to this matrix as the *finite-field transform matrix*. The form of this matrix is
very similar to that of the discrete-Fourier-transform matrix in complex numbers, and
some of the well-known properties of this transform are preserved in the finite field.
Note that the last row of the matrix can be interpreted as a list of the inverses of the
elements in the second row.

If the information vector is extended by zeros to a vector of length n, we can write
the encoding of the RS code as

$$c = uF,$$

thus interpreting the encoding as a finite-field transform. The generator matrix, G, is the
first k rows of F.

Example 4.5 (Encoding in $F(11)$). *In $F(11)$ 2 is a primitive element. The encoding can be expressed as the following finite transform:*

$$c = [3\,2\,0\,0\,1\,0\,0\,0\,0\,0] \begin{bmatrix} 1 & 1 & 1 & 1 & 1 & 1 & 1 & 1 & 1 & 1 \\ 1 & 2 & 4 & 8 & 5 & 10 & 9 & 7 & 3 & 6 \\ 1 & 4 & 5 & 9 & 3 & 1 & 4 & 5 & 9 & 3 \\ 1 & 8 & 9 & 6 & 4 & 10 & 3 & 2 & 5 & 7 \\ 1 & 5 & 3 & 4 & 9 & 1 & 5 & 3 & 4 & 9 \\ 1 & 10 & 1 & 10 & 1 & 10 & 1 & 10 & 1 & 10 \\ 1 & 9 & 4 & 3 & 5 & 1 & 9 & 4 & 3 & 5 \\ 1 & 7 & 5 & 2 & 3 & 10 & 4 & 6 & 9 & 8 \\ 1 & 3 & 9 & 5 & 4 & 1 & 3 & 9 & 5 & 4 \\ 1 & 6 & 3 & 7 & 9 & 10 & 5 & 8 & 4 & 2 \end{bmatrix}$$

$$c = [6\,1\,3\,1\,0\,2\,4\,9\,2\,2].$$

The inverse of the transform matrix, F, is

$$F^{-1} = -(\alpha^{-ij}).$$

Thus, as in the complex field, the inverse transform differs from the original transform only by a reordering of the rows and a scaling factor. One consequence is that we can find the parity-check matrix, H, as the last $n - k$ rows of the inverse matrix, or we can take them from the top of F starting with row 2.

In the definition of the RS codes we used the interpretation of the information vector as a polynomial, and the encoding was performed by evaluating this polynomial. We can now take the opposite approach, reading the codeword as the polynomial

$$c(x) = c_{n-1}x^{n-1} + \cdots + c_1x + c_0.$$

Since the inverse transform must produce zeros for the first $n - k$ powers of α, these elements are zeros of all codewords, and any polynomial that has zeros at these points is a codeword. Consequently we can characterize codewords in the following way: the codewords of an (n, k) RS code have the generator polynomial

$$g(x) = (x - \alpha)(x - \alpha^2)\ldots(x - \alpha^{n-k})$$

as a factor. Thus, with the ordering of the symbols, an RS code is cyclic, as discussed in Chapter 3.

We could perform a valid encoding of an information polynomial of degree less than k by multiplying it by g; however, the preferred encoding is to use the polynomial

$$c(x) = u(x)x^{n-k} - (u(x)x^{n-k} \bmod g(x)).$$

Thus the information is shifted to the left, and then divided by g, and the remainder is appended at the right. The reason for this approach is that the information symbols appear directly in the codeword, which is referred to as *systematic encoding*.

One reason for this choice is that, even if decoding fails, the information can be recovered with a few errors. The encoding used in the previous section is called *non-systematic*.

In the coefficient matrix that we used to find Q the left part is equal to a segment of F. Multiplying by the rows of F that are orthogonal to this part eliminates Q_0, and we get a system of equations for Q_1. The corresponding matrix is conveniently written using the syndromes

$$S_i = c(\alpha^i),$$

$$\begin{bmatrix} S_1 & S_2 & \cdots & S_{t+1} \\ S_2 & S_3 & \cdots & S_{t+2} \\ \vdots & \vdots & \ddots & \vdots \\ S_t & S_{t+1} & \cdots & S_{2t} \end{bmatrix} \begin{bmatrix} Q_{1,0} \\ Q_{1,1} \\ \vdots \\ Q_{1,t} \end{bmatrix} = \begin{bmatrix} 0 \\ 0 \\ \vdots \\ 0 \end{bmatrix}. \tag{4.4}$$

Note that, while we have earlier used the term syndrome to refer to the vector Hr', it is conventional in discussing RS codes to use the term for the individual entries of the vector.

Example 4.6 (Decoding RS codes from syndromes).   *As in the previous example we decode the $(10, 5)$ code over $F(11)$. If the received vector is r, the syndromes are found from the transform, and the error locator is obtained from the syndrome matrix. The calculation is performed in the following MATLAB script.*

```
%Decoding from syndromes
p=11;                   %The symbol field is integers modulo p
n=p-1;                  %length of code
t=2;                    %number of errors corrected

%Set up transform matrix

ff1=mod(2.^[0:1:p-2],p);     %powers of primitive element
iin1=[0:p-2];                %indices
iin2=mod(iin1'*iin1,p-1);    %exponents for transform matrix
ff2=mod(2.^iin2,p);          %transform matrix

%Encode by transform

uu=[3 2 0 0 1 0 0 0 0 0];    %information
cc=mod(ff2*uu',p)';          %encoding as in Example 4.5
%but the encoded symbols are in a different order
ee=[1 2 0 0 0 0 0 0 0 0];    %error pattern
rr=mod(cc+ee,p);             %received word with errors
```

%Find the error locator from syndromes

```
ss=mod(ff2*rr',p)';          %syndromes are ss(2:7)
ssm=[ss(2:4);ss(3:5);ss(4:6)]; %syndrome matrix
q1=mod([2 -3 1],p);          %error locator
mod(ssm*q1',p)'              %verify that (4.4) is satisfied
```

Since the matrices involved in decoding RS codes have a particularly simple structure, special methods can be used for solving these systems. One such algorithm will be presented after we have developed the necessary tools for doing computations in $F(2^m)$.

4.6   Reed–Solomon codes over the fields $F(2^m)$

Since most applications of RS codes use fields with 2^m elements to encode data consisting of m-bit words (in most cases $m = 8$), we provide enough background in this section to allow us to discuss decoding in such (extension) fields. The integers with addition and multiplication performed modulo 2^m are *not* a field, since certain products of nonzero elements give a zero result, and thus such elements do not have inverses. Instead we interpret each m-bit sequence as a polynomial of degree less than m with coefficients in the field $F(2)$. The usual convention is that the leftmost bit is the highest-order coefficient.

Addition is performed position-wise (like addition of polynomials), and such an *xor* function on words is commonly available both in programming languages and in digital logic. Thus there are no carries, and, for any element in the field, $a + a = 0$, i.e. subtraction is the same as addition.

In order to specify multiplication in a way that gives a result in the field, we need to reduce the product of two polynomials modulo a polynomial of degree m. Thus the field is specified in terms of a polynomial, $p(z)$, which must be chosen to be irreducible, i.e. it does not factor into polynomials of lower degree with coefficients in $F(2)$. We use a different indeterminate, z, to avoid confusion with the polynomial representation of the codeword. This polynomial plays a role similar to the prime that defines $F(p)$. The existence of irreducible polynomials can be proved by noting that there are more polynomials of a given degree than products of lower-degree irreducible polynomials. Usually a suitable irreducible polynomial is selected from available tables. One can prove that two fields specified by irreducible polynomials of the same degree are equivalent, so it is not very important which one is chosen.

It is often useful to have a primitive element available. By a proper choice of $p(z)$ (as a so-called primitive polynomial), we can make the element z primitive. Thus, starting from 1 and in each step multiplying by z modulo $p(z)$, we will get all nonzero polynomials of degree less than m, and eventually 1 is obtained in step $2^m - 1$. We may think of this process as shifting the current string to the left, and, if the leading coefficient is 1, adding $P(z)$. This way of listing the nonzero strings was mentioned earlier as a way of making Hamming codes cyclic. Once each element of the field is known as a power of the primitive element, $a_j = z^j$ modulo $p(z)$, we can multiply two nonzero elements by adding their exponents modulo $2^m - 1$. If one of the factors is zero, the product is always zero. Multiplication is usually performed by using a table of exponents or a full multiplication table.

The existence of inverses is proved as for prime fields, and division can be performed using a table of inverses or by subtracting the exponents modulo $2^m - 1$.

We can now construct, encode, and decode RS codes using the arithmetic of $F(2^m)$.

Example 4.7 (RS codes over $F(16)$).   *The field $F(16)$ can be constructed using the primitive polynomial $p(z) = z^4 + z + 1$. Thus $\alpha = z$ is the primitive element, and we find $\alpha^2 = z^2$, $\alpha^4 = z + 1$, $\alpha^8 = z^2 + 1$, and $\alpha^{16} = z = \alpha$. A $(15, 11, 5)$ RS code correcting two errors has generator polynomial*

$$g(x) = (x + \alpha)(x + \alpha^2)(x + \alpha^3)(x + \alpha^4)$$
$$= x^4 + (1 + z^2 + z^3)x^3 + (z + z^3)x^2 + z^3 x + (1 + z + z^2)$$
$$= x^4 + \alpha^7 x^3 + \alpha^9 x^2 + \alpha^3 x + \alpha^{10}.$$

A convenient notation for doing calculations in the field is to represent polynomials in x as vectors and convert the field elements into numbers between 0 and 15 in binary notation. Thus we can write the generator polynomial as $(1, 13, 10, 8, 7)$.

The details of encoding and decoding an RS code in $F(16)$ are provided as a MATLAB script in Appendix C.

4.7   Faster decoding methods

The decoding methods described earlier require the solution of systems of linear equations. While such calculations are straightforward in finite fields (numerical stability is not a concern), it is often desirable to reduce the amount of computation by exploiting the special structure of the coefficient matrices. Here we give an outline of one such algorithm that is commonly used. The syndromes $S_j = r(\alpha^j)$ are calculated and interpreted as the syndrome polynomial

$$S(x) = S_1 x^{2t-1} + S_2 x^{2t-2} + \cdots + S_{2t}.$$

We then perform part of the *extended Euclid algorithm* on the polynomials x^{2t} and $S(x)$:

(i) $a_{-1} = x^{2t}$, $a_0 = S(x)$, $q_{-1} = 0$, $q_0 = 1$;
(ii) for $j = 0 \ldots$ until $\deg(a_j) < t$;
(iii) $a_{j-1} = a_j b_j + a_{j+1}$, $\deg(a_j) < \deg(a_{j-1})$;
(iv) $q_{j-1} = q_j b_j + q_{j+1}$.

The last $q(x)$ is then the error locator, and the remainder a of degree less than t is called the error-evaluator polynomial. (Euclid's algorithm is meant to find the greatest common divisor of the two input polynomials, and runs until the remainder is zero or a constant. However, here it is stopped about halfway through).

We omit a detailed derivation of the algorithm, but the rationale can be explained in the following way. We can think of the sequence of syndromes as the first terms in a power series for a rational function. In each step of the algorithm, the polynomials satisfy

$$q_j(x)S(x) = a_j(x) \bmod x^{2t}. \tag{4.5}$$

In particular the error-locator and -evaluator polynomials satisfy this relation. We can express $S(x)$ by the first $2t$ terms in an expansion of the rational function, a/q.

We then find the error locations as the roots of the error locator, x_j. The number of roots should equal the degree of the polynomial; if this is not the case, the received vector contains more than t errors. The contributions from each error can be expressed as

$$e_j x_j^{2t} + e_j x_j^{2t-1} x + \cdots = \frac{e_j x_j^{2t}}{1 - x/x_j}.$$

We can now rewrite the syndrome sequence by decomposing the rational function into partial fractions:

$$\frac{a}{q} = \sum_j \frac{e_j}{x - x_j}.$$

Here q factors into terms corresponding to the error positions, and the numerators of the partial fractions are the error values. The last step of the algorithm uses a well-known method for splitting a rational function into partial fractions (or one can say that it uses an explicit formula for the inverse of the matrix in question):

$$e_j = -x_j^{2t+1} \frac{a(x_j)}{q'(x_j)}. \tag{4.6}$$

Here q' indicates the derivative of q. For each term x^j, the derivative in a prime field is jx^{j-1} and that in $F(2^m)$ is simply x^{j-1} for odd j and zero for even j.

Further details are given in Appendix C.

Decoders are often implemented in programmable-logic arrays, and it is important that much of the computation can be performed with a high degree of parallel processing: the syndromes are computed in parallel, and the scaling of polynomials in Euclid's algorithm and in the search for roots in the error locator can similarly be implemented in parallel.

In principle, even faster computations could be obtained by evaluating the syndromes as a fast transform of the received sequence (using fast-Fourier-transform methods). Similarly, the error positions could be obtained as a fast transform.

Figure 4.1. A Datamatrix encoding the title of this book.

Example 4.8 (Datamatrix code).   *There are several 2-D versions of barcodes for product identification. One commonly used format is the Datamatrix code, which is protected by an RS code. Figure 4.1 shows an example in the form of a 26 by 26 square. The left and lower boundaries are always black, while the other two boundaries alternate between black and white. The remaining 576 bits contain 72 symbols of an RS code over $F(256)$. Most of the symbols are mapped into 3 by 3 squares with the upper-right bits cut out. These symbols are stacked along diagonals. Most of the remaining area along the boundaries is used for additional symbols.*

4.8   Reed–Solomon codes and q-ary channels

As discussed in Chapter 3, a symmetric q-ary channel can be defined by an error probability and the assumption that, when an error occurs, all of the other $q - 1$ symbols appear with equal probability. Following the discussion of the capacity of the symmetric q-ary channel, we can say that t syndromes are required for the calculation of error values, and that in the decoding of RS codes another t syndromes are required for the calculation of the positions. The first number is close to the minimum, whereas the second number differs from the bound by a factor $t!$. For large N and small t, it is not possible to correct more than t errors with the $2t$ available syndromes, whereas for low rates it would be possible to correct a significantly larger number of errors.

   If the actual error values are far from being equally distributed, it is also possible to correct more errors. Thus, if there are few bit errors in the erroneous symbols, or if the symbols are associated with positions in the plane such that each symbol has a small number of neighbors (as in some modulation formats), the channel has a higher capacity for a given error probability. However, in applications such as printed labels, the errors may be far from evenly distributed, but it is not easy to provide a likely distribution.

4.9 Notes

The Datamatrix 2-D barcode is described in the standard ISO/IEC 16022, of the International Organization for Standardization (2006).

Exercises

4.1 For $p = 17$, construct tables for addition and multiplication in $F(p)$. Show that 3 is a primitive element, but 2 is not.

4.2 Find a generator matrix for the code in Example 4.1. Decode the received vector (0 0 0 1 1 0 0 2 3 5).

4.3 Encode the information sequence (0 0 0 0 0 0 1 0 3 2) using a (16, 10) RS code over $F(17)$. How many errors can the code correct?

4.4 Use the script in Example 4.4 to decode the $F(17)$ code with various error patterns.

4.5 Find a generator polynomial for the $F(17)$ code.

4.6 Use syndrome decoding to decode several error patterns with the $F(17)$ code.

4.7 F(256) can be constructed using the primitive polynomial $p(x) = x^8 + x^5 + x^2 + x + 1$. Generate a table of exponents. Multiply the two elements $x^5 + 1$ and $x^4 + x + 1$ using the exponent table, and compare the result with that obtained by a direct multiplication and reduction modulo $p(x)$.

4.8 Decode a single error in the $F(16)$ code without using the examples.

5   Source coding

This chapter treats source coding, with a focus on the basics of lossless coding of image and graphic data. Data compression and image coding are widely used today when transmitting and storing data. Examples are transmission of images on the Internet and storage of images, audio files, and video on CDs or DVDs.

In Chapter 1 source coding of discrete sources generating independent symbols was introduced, specifically in the form of Huffman codes and arithmetic coding. In Chapter 2 some source models were considered. In this chapter we treat source coding of statistically dependent data. Furthermore, for real-world data the model is unknown, so this chapter also deals with issues concerning efficiently estimating model parameters and possibly the model order for source coding. Actually, the standard notion of a model may be abandoned altogether and the goal may be to code a single individual sequence, i.e. independently of other sequences and their statistics.

First we consider how to code the source data in relation to the models previously defined. In general terms, the model is represented by a context defining a dependency on the past data.

We shall refer to the approach as context-adaptive coding. The context may readily be defined in two or more dimensions. Doing so naturally leads to source coding of images. This forms the basis of the binary image-coding standards originally developed for fax coding and today used in a number of applications including the widely used PDF (Portable Document Format) format. This provides lossless coding of the binary images, such that the decoder can reconstruct exactly the same binary data as the encoder coded.

For lossless gray-level or color image coding, a prediction or transformation step is introduced, but the final entropy coding is also based on context-adaptive coding. The context specifies the relations to the past in the form of conditional probabilities. The power of the approach is that the context may represent two-dimensional (2-D) structures (or in general any structures), while the coding is still performed in a sequential order to produce a code string.

For image and video coding higher compression may be achieved if a loss is accepted. Well-known standards are JPEG, JPEG2000 for images, and MPEG for video. In these lossy coding schemes the data are first processed and then a quantization step is introduced to reduce the code length at the expense of distortion with respect to the input data. Again entropy coding is applied, and, for the most efficient of these standards, it

Figure 5.1. A 2-D context defined by the four causal neighbors.

is based on the techniques of context-adaptive coding introduced in this chapter. Image coding is treated in the next chapter.

5.1 Context-based adaptive coding of sources with memory

Markov chains were presented in Chapter 2 as data sources having a memory, which is specified by the transition probability, $p(x_t|x_{t-1})$. Markov sources have a finite memory given by the conditional probability $p(x_t|x_{t-m}^{t-1})$ based on m previous symbols, $x_{t-m}, \ldots, x_{t-1}$. In two dimensions, the Pickard random field was presented on the basis of the conditional probability $P(D|ABC)$. These sources are all characterized by the Markov property, i.e. the conditional probability based on a finite part of the past defines the model. For Markov sources the finite memory may be represented in terms of a state, such that the new state after each transition is given by the previous state and the output symbol of the transition. This is not the case for the Pickard model. We refer to the neighborhood defining the conditional probability as the context, which is more general than the notion of a state. For a given model and the conditional probabilities, the entropy may be determined by virtue of the chain rule. The notion of a context is further generalized to become (a mapping of) the previous symbols used when coding the next symbol. By the same argument optimal source coding may be performed sequentially one element at a time conditioned on the context.

Formally, arithmetic coding may take a sequence of symbols, x_1^T, and corresponding probability assignments, $P(x_t|x^{t-1})$, and code the symbols, such that the total code length is very close to

$$L(x^T) = -\sum_{t=1}^{T} \log P(x_t|x^{t-1}), \qquad (5.1)$$

which is called the *ideal adaptive code length*.

This has led to the modern paradigm of data compression separating the coding into a modelling step to attain $P(x_t|x^{t-1})$ and a coding step based on arithmetic coding. The interval subdivision in arithmetic coding (see Chapter 1) corresponds to the contribution $-\log P(x_t|x^{t-1})$ to the ideal adaptive code length. We shall pursue this and consider the modelling step, assuming that arithmetic coding is used subsequently.

For real data, probabilities are at best unknown and it may even be questionable whether they exist. So the modelling step becomes a question of determining both a model within a model class and the parameters of the model. For Markov sources the parameters are the conditional probabilities. We shall consider this issue for a single data set $x = x^T$, which enables us to do the coding regardless of the question concerning the existence of the model.

5.1.1   Two-part coding

The objective is to code a data set x^T with elements drawn from a finite alphabet $\mathcal{A}$. Let $|\mathcal{A}|$ denote the size of the alphabet. Assuming a model, the parameters may be determined on the basis of occurrences within the data set. Thereafter the model parameters so defined may be coded as a preamble and put in the header. Finally the entropy coding with the given parameters may be performed. This can be extended and the encoder could try all models within the class and choose the model giving the minimum code length (including the header). If we consider the Markov source, the model could be specified by the memory m in $p(x_t|x_{t-m}^{t-1})$ and the parameters are the conditional probabilities. All values of m could be probed up to some maximum number. This gives $(|\mathcal{A}| - 1)|\mathcal{A}|^m$ parameters (plus coding the memory order, m). One instance of this is determining a Huffman code and coding it as part of the header. This is often referred to as a *Huffman table* and the approach may be used in e.g. JPEG image and MPEG video coding. Assuming independence, i.e. that there is no memory of the symbols to be coded, one table is sufficient. For the general Markov source, $|\mathcal{A}|^m$ tables are required. This gives rise to a number of issues, e.g. there is an initial pass of the data (and thereby latency) in order to acquire the statistics, there is the question of how model information should efficiently be coded, and, related to this, how do we start? There are solutions to these problems, but we shall instead focus on the adaptive approach, in which the model parameters and possibly also the model order are learned as we proceed.

5.1.2   Context-adaptive code length

First consider a sequence of independent binary variables, X_t, taken from the same distribution, i.e. an i.i.d. process. The objective is to estimate the probability at time t given the past x^{t-1}. Consider k observations from the sequence. Let $n_0(k)$ denote the number of 0s and $n_1(k)$ the number of 1s out of the k symbols. A simple estimate is given by

$$\hat{p}_1 = \frac{n_1(k) + \delta}{n_1(k) + n_0(k) + 2\delta}. \tag{5.2}$$

For the two-part coding above, we would select $\delta = 0$ to minimize the code length. In the adaptive setting we cannot fully rely on the occurrence counts, but a small value of δ is selected; $\delta = 1$ gives integer counts, whereas $\delta = 1/2$ has good theoretical justification. The probability estimate is a mixture of the occurrence-count estimate and the uniform distribution represented by δ. Obviously, the estimate $\hat{p}_0$, of the probability of a 0 is obtained by exchanging $n_0(k)$ and $n_1(k)$ in (5.2).

The probability estimate may be seen as a Bayes estimate. If the probability p_1 itself is considered a stochastic variable with a beta distribution ($p_1 \in B(\delta, \delta)$) specified by the parameter δ, the Bayes estimate of p_1 given k observations is given by (5.2).

$\delta = 1$ is optimal if the prior distribution of p_1 is uniform. Smaller values of δ reflect an expectation of one of the symbols being more probable, but without knowledge of which beforehand ($\delta = 0.45$ is used in the JBIG bi-level image-coding standard).

Introducing contexts, the estimate based on $n_0(k)$ and $n_1(k)$ is used for each context and k refers to the number of occurrences of the given context in the string x^t. This estimate is used when the context appears the next time, i.e. at $k + 1$.

Now we formulate this in the general setting with a finite alphabet, $\mathcal{A}$. A probability is assigned by $P(x^T) = \prod P(x_t|x^{t-1})$, where $P(x_t|x^{t-1})$ is expressed by $P(x_t|x^{t-1}) = P(x_t|F(x^{t-1}))$. The *context function* $F(x^{t-1})$ defines the context in terms of a mapping of the causal elements. In context-based adaptive coding the probability assignment $P(x_t|x^{t-1})$ is updated sequentially. A probability assignment based on occurrence counts in each context is considered. Let $r_t = F(x^{t-1})$ be the context value at t and $n_t(a|r_t)$ be the number of times that value a appears with context r_t in the sequence x^{t-1}. The probability assignment (5.2), for $\delta = 1/2$, is

$$P(a|r_t) = \frac{n_t(a|r_t) + 1/2}{\sum\limits_{x \in \mathcal{A}} n_t(x|r_t) + |\mathcal{A}|/2}. \tag{5.3}$$

The ideal code length for context-adaptive coding is $-\sum_t \log P(x_t|r_t)$ and the contribution to the ideal code length of x^T by the occurrences in a specific context denoted s is given by

$$L(s) = -\sum\limits_{t|r_t=s} \log P(x_t|r_t = s), \tag{5.4}$$

where $P(x_t|r_t = s)$ is given by (5.3).

The ideal code length, $L(x^t)$, is given by summing $L(s)$ over all contexts,

$$L(x^t) = \sum\limits_{s} L(s). \tag{5.5}$$

Using arithmetic coding, we may come very close to this value. *Context-based adaptive arithmetic coding* may be expressed in three steps. For each symbol x_t, it

(i) determines the context $r_t = F(x^{t-1})$ and the conditional probability $P(x_t|r_t)$ based on occurrence counts (5.3),
(ii) performs arithmetic coding of x_t based on $P(x_t|r_t)$ (Chapter 2), and
(iii) updates the occurrence counts for context r_t by incrementing $n(x_t = a|r_t = s)$.

On taking a closer look at $L(s)$ and the underlying probability estimate (5.3), it is seen that the adaptive code length depends only on the set of counts $\{n_T(a|s)\}$ and not on the order in which they appear. Thus the ideal code lengths may be expressed just on the basis of the occurrence counts.

Without loss of generality, we consider the i.i.d. binary case with counts (n_0, n_1) and $n = n_0 + n_1$. Using the estimate (5.2) sequentially with $\delta = 1$ gives the adaptive code length

$$L(s) = \sum_{i=1}^{n} \log(i+1) - \sum_{j=1}^{n_0} \log j - \sum_{k=1}^{n_1} \log k$$

$$= \log(n+1)! - \log n_0! - \log n_1!. \tag{5.6}$$

Stirling's approximation may applied to this expression. One version is given by

$$\log n! = (n+1/2)\log n - n + \ln\sqrt{2\pi} + R(n), \tag{5.7}$$

where $1/(12(n+1)) \le R(n) \le 1/(12n)$.

Applying Stirling's approximation to (5.6) gives $nH(n_0/n)$ as the leading term. (This is consistent with the approximation given in Chapter 3.) The remaining parts may be perceived as the model cost, i.e. the price for adapting to the statistics. This model cost is on the order of $1/2\log n$. In the binary case above there is just one parameter. This may be generalized such that, for K parameters, the model cost is $K/2\log n$. This term also clearly shows that, for a given sequence of length n, the model should not have too many parameters, which is referred to as the context-dilution problem. If different models are to be evaluated in terms of the ideal code length, the term $\log n!$ may be tabelized for fast calculation of code length.

The cost of coding symbols on the basis of a distribution that does not match the statistics of the data may also be expressed as follows. Consider coding x having probability $P(x)$ assuming a different distribution that assigns the probability $Q(x)$. The expected code length will be $-\sum P(x)\log Q(x)$. Subtracting the entropy yields a redundancy of

$$\sum P(x)\log\left(\frac{P(x)}{Q(x)}\right),$$

which is also referred to as the *divergence* $D(P||Q)$ of the distributions $P(x)$ and $Q(x)$. For a given code, the term $-\log Q(x)$ may be replaced by the number of bits used to code x.

5.1.3   Context-based adaptive image coding

The context approach introduced above defining the dependency on the past provides an efficient way of utilizing 2-D (and higher-dimensional) dependencies. Let x_{ij} denote the picture element at (i, j), drawn from a finite alphabet, $\mathcal{A}$. For compression, the images are processed sequentially. The pixel position (i, j) is mapped onto a sequence index t by a traversal of the image. Let x_t denote element t in the sequential representation of the image and x^T the whole image. We shall use the conventional row-by-row raster scan. The example below illustrates this for the Pickard random fields.

Example 5.1 (Adaptive context-based image coding of a Pickard random field).   *For the Pickard model (Chapter 2), the conditional probabilities are given by $P(D|ABC)$ (apart from the first row and the first column, which we assume given for simplicity). Now consider a coding model described by the conditional probabilities $P(d|abc)$ for* *all values of a, b, c, d, but assume that these probabilities are not known. To code a given image, this is traversed (row by row) and coded sequentially one symbol at a time. The conditional probability for each value of $d \in \mathcal{A}$, given the context $abc \in \mathcal{A}^3$, may be estimated by (5.3) on the basis of the occurrences so far. The sequence of symbols is coded on the basis of these conditional probabilities by arithmetic coding. Here the coding is based on estimating $P(D|ABC)$, without strictly imposing the extra constraints of the Pickard random field (PRF), e.g. $P(D|ABC)$ is not restricted by requiring consistency with a distribution on (ABC) that satisfies the PRF independence condition. Thus the scheme may be applied as an approximation to a broader class of images.*

To code an image with T elements in sequential order expressed by x_t, a probability is assigned by $P(x^T) = \prod P(x_t|x^{t-1})$, where $P(x_t|x^{t-1})$ is expressed by $P(x_t|x^{t-1}) = P(x_t|F(x^{t-1}))$.

We shall mainly consider context functions, $F(x^{t-1})$, defined by taking a subset of elements of the past with given position relative to the current element, $x_t = x_{ij}$, as in the Pickard example. This is referred to as *template-based coding*. For a sequence of variables $X_1^t = X_1, \ldots, X_t$, a template-based context with K elements is expressed for each position, t, by a subset of the previous variables, $(X_{t-t_1}, X_{t-t_2}, \ldots, X_{t-t_K})$. (Again we disregard the boundary for 2-D data. In this case a special mapping shall be defined.)

Arithmetic coding may be applied sequentially to the conditional probabilities. Thus the definition of the context function $F(x^{t-1})$ becomes crucial in the design of a context-based lossless compression algorithm.

5.1.4 Context-based coding of binary images

The template-based coding presented above may be applied to the coding of binary images, achieving very efficient compression.

Given an image, the choice of template size and pixel locations may be determined by searching. A simple greedy search often provides good results (for binary images), but more advanced searching may also be applied. For domain-specific applications a predefined template will often provide good results, as reflected by the standards. Arithmetic coding may be applied to the conditional probabilities. In many practical applications, a fast version of arithmetic coding is desired.

Example 5.2 (Bi-level template-based image coding – JBIG). *A template of ten pixels is used in the bi-level image-coding international (ISO) standard JBIG. The position of one of these may be chosen by the encoder. In the follow-up standard, JBIG2, up to 16 template pixels may be used and the positions of up to four of these may be specified by the encoder. JBIG2 also specifies other techniques, which shall be presented in the next chapter. A major application of JBIG is coding of documents in raster representation.*

In the JBIG standards, multiplications and divisions both of the arithmetic coding and of the conditional probabilities are avoided. An approximative solution avoiding

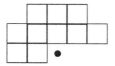

Figure 5.2. A ten-pixel template as used in JBIG.

multiplications is used to speed up the basic recursions. Consider the binary arithmetic coder and let q denote the smallest of the two probabilities, i.e. $q = \min\{P(0|x^n), P(1|x^n)\}$. The interval shall be represented by the lower bound on the interval, l_n, corresponding to $C(x^n)$ and the interval width $A(x^n)$. In each step, the interval is scaled by powers of 2 such that $3/4 \leq A(x^n) < 3/2$. By approximating $A(x^n)q \approx q$ the recursion is simplified and multiplications and divisions avoided. Details are given in Appendix A. The probabilities assigned for coding are based on (5.3), but, to avoid the need for division, the adaptive probability estimates are implemented using a finite-state machine in an approximating form for speed in the JBIG standards. More details are given in Appendix A.

The principles above leading to multiplication-free fast arithmetic coding (in JBIG and JBIG2) are also used in the entropy coding of other contemporary standards for image and video coding: for gray-scale and color images (JPEG2000, JPEG-LS) and video (MPEG4/H.264). (In JPEG-LS and MPEG4/H.264 the user application may choose between Huffman-like tabelized variable-length coding (VLC) and adaptive binary arithmetic coding.) The variations of arithmetic coding used are further developments of the arithmetic coder outlined in the example above and, in more detail, in Appendix A.

5.2   Universal coding

So far the model order has been fixed. In the previous section template-based coding was considered, in which the context function is fixed, but the probability values are learned adaptively (5.3). For real data the (best) model will not be known beforehand. Given a finite number of models, a data set could be coded with all of these and the best model chosen. In this section we shall consider the more elegant alternative of double adaptive coding, whereby both the model (within a class) and the model parameters are determined adaptively as we proceed. The goal is to do so with limited or minimal loss of coding performance. This goal may be expressed by the term *universal coding*, which requires that the per-symbol code length asymptotically converges to the entropy of the source.

We shall present two algorithms providing universal coding. The *Lempel–Ziv (LZ)* schemes are used in popular file-compression systems. Algorithm Context has better performance but is also more complex. The better performance is reflected theoretically by a faster convergence rate than that of the LZ codes.

5.2.1   Lempel–Ziv coding

The file-compression systems of today are based on one of two universal coding schemes invented by J. Ziv and A. Lempel and referred to as LZ coding. Files in general may be generated in a variety of ways and representing many different sources. Thus a compression scheme that can learn the structure is desirable.

Huffman coding may be used as a fixed-to-variable-length code, where a fixed number of source symbols with probability p is assigned a code word with a length of approximately $-\log p$. This approach may be turned around to give a variable-to-fixed-length code, in which the source sequence is parsed to segments having probability $p \approx 2^l$.

There are two basic versions of LZ universal coding. The first (LZ1 or LZ77) is based on finding the longest match of the symbols to be coded within a window of the past symbols. Here the sequences of symbols present in the causal part of the string may be seen as a dictionary, which is searched in order to find a long match. The longest match is coded by an offset value into the past and the length of the match. In the original version the next character was also coded. The other method (LZ2 or LZ78) is based on explicitly building a dictionary. In both cases the matches are given by strings of consecutive characters, which do not need to be complete words. The second version is presented in more detail below.

5.2.1.1   The LZ78 principle

The LZ78 (or LZ2) dictionary-based coding consists of adaptively building a dictionary and, for each new coding step, finding the longest match of the coming symbols in the dictionary. Consider a string of symbols, x_1^T, drawn from the alphabet $\mathcal{A}$. In the original LZ78 version, for each coding step, the index i of the longest match is coded together with the next symbol, x, coded by $C(x)$. The combined codeword, which is denoted $\langle i, C(x) \rangle$, is transmitted and the substring given by the combined string is inserted as the next word in the dictionary both at the encoder side and at the decoder side. To formalize the description, the actions when coding a new substring starting at t are as follows.

(1) Find the longest match in the dictionary, i.e. the largest value of k for which x_t^{t+k} matches an entry in the dictionary.
(2) Code the index i of this longest match and the next character $C(x)$, $\langle i, C(x) \rangle$.
(3) Insert the substring x_t^{t+k+1}, given by the concatenation of the longest match, x_t^{t+k}, and the following character, x_{t+k+1}, as a new symbol with index K in the dictionary and increment K.

Having K entries in the dictionary, a simple fixed-length binary representation will use $\lceil \log K \rceil + \lceil \log|\mathcal{A}| \rceil$ bits per string. Viewing the dictionary as a tree, it will grow faster along frequently occurring paths of this tree, thus balancing the probabilities of the words in the dictionary, which are leaves on the tree.

Example 5.3 (Lempel–Ziv coding). *Consider applying LZ78 to a string of binary symbols, $x_1^{10} = 0101010101$. Initially the dictionary is empty. The string is parsed as $0|1|0,1|01,0|1,0|10,1$, where "|" delimits the entries into the dictionary and "," delimits*

Table 5.1. Lempel–Ziv code and dictionary development

Output	Entry	Index (i)
0, $C(0)$	0	1
0, $C(1)$	1	2
1, $C(1)$	01	3
3, $C(0)$	010	4
2, $C(0)$	10	5
5, $C(1)$	101	6

the longest match and the following symbol. The two first entries have the empty string as longest match. The output of the encoder and the dictionary after the string is given in Table 5.1.

To decode the code string generated, the decoder just has to update the dictionary each time it receives $\langle i, C(x) \rangle$, in order for it to be synchronized with the dynamic dictionary at the encoder side.

It may be noticed that the next character represented by $C(x)$ is not compressed. This has no significance asymptotically, but, for finite-length messages, it does lead to a loss of compression. A remedy is simply obtained by not coding the next character in Step 1, but in Step 2 reading $C(x)$ as the first character of the next string. This variation is attributed to Welch and it is referred to as LZW coding. In this version, the dictionary is initialized with the symbols of the alphabet, such that only an index is sent for each substring including the initial phase. Each time a new symbol occurs in the data string, this leads to just this one symbol being coded by the next codeword. To facilitate implementation, a maximum size was imposed on the dictionary. This maximum size may be parameterized within some limits.

Example 5.4 (LZW coding). *Consider applying LZW coding to the string of binary symbols in Example 5.3, $x_1^{10} = 0101010101$. The dictionary is primed with an entry representing each of the symbols of the alphabet, here 0 and 1. The string is parsed as 0,1,01,01,010,101, where "," delimits the longest match. The new entry is given by the longest match and the next symbol to be read from the next match. The output of the encoder and the dictionary development for the string, x_1^{10}, are given in Table 5.2. (Each new entry is not completed at the decoder side until the first symbol in the next match is read, but this does not cause a conflict because it will be available when needed even if the longest match refers to this latest entry.)*

While LZW solves the problem of having uncoded symbols, it still has the drawback of disregarding the correlation across segments that are parsed to different phrases.

LZW is used in a number of well-known file-compression schemes. The Unix compress *command was probably the first widely used scheme. In* compress *the scheme is byte-oriented, i.e. defined over an alphabet of size 256. This is usually the case for*

Table 5.2. LZW code and dictionary development

Output	Entry	Index (i)
	0	1
	1	2
1	01	3
2	10	4
3	010	5
5	0101	6
4	101	7

general file compression, but may be changed for more specific applications. It is natural to try to match the alphabet of the LZ coder with that of the data.

Two widely used coding schemes for graphics, GIF and PNG, are based on each of the basic LZ coding schemes. They may be applied to pixel-based graphics, i.e. when the file is represented as a discrete image.

Example 5.5 (Coding of graphics based on LZ coding).   *The Graphics Interchange Format (GIF), which is based on LZW coding, is widely used for coding graphics, e.g. for internet applications. The discrete image is converted into a string traversing the image row by row. Thereafter LZW coding is applied to this string, defining the alphabet as that of the image, if this does not exceed 8 bpp (bits per pixel). Color images and gray-scale images having more than 8-bpp are mapped to an 8-bit index, i.e. quantized to 8 bpp prior to the LZW coding. The drawback when coding graphics using GIF is that the 2-D structure is not captured. An alternative LZ-based coding scheme for graphics and images is PNG (Portable Network Graphics). This is based on zip, which in turn is based on LZ1. In PNG, the 2-D structure of the image data may to some extent be captured by predictive prefiltering, e.g. by subtracting the value of the pixel above and then LZ coding the resulting difference or residual. Since these residuals may assume both positive and negative values, they are without loss represented modulo 256 to maintain an 8-bit representation. For images this predictive filtering will in general improve compression, but the 2-D modelling is very simple, meaning that coding based on better 2-D models may lead to better compression. Especially for graphics, the simple predictive filter may be far from an optimal model and context-based coding may provide better results.*

The widely used zip compression is based on LZ1. It will in general provide better performance than do LZW schemes. The advantage from a practical point of view of LZ codes compared with, e.g., context-based methods, is that they are coding multiple characters fast in each step. In their basic version they do converge to the rate given by the entropy for finite-memory sources, thus providing universal coding. An inherent

drawback is that, when parsing a sequence into substrings and coding these, the corre-
lations between these substrings are not utilized. From a theoretical point of view, this
leads to a slower convergence to the entropy rate, which in turn may lead to inferior
compression performance for finite data sets.

5.2.2　Algorithm Context

An elegant solution to universal coding is provided by Algorithm Context. The template
introduced previously to define a fixed-length context is replaced by a dynamic tree
structure, where also the context length is decided adaptively. Consider a sequence of
symbols, x^T, drawn from a finite alphabet. Given an (unbounded) ordering of context
elements, $(X_{t-t_1}, X_{t-t_2}, \ldots)$, where each t_i represents the relative displacement of a
context pixel. For each new element, x_t, the algorithm consists of two steps.

(i) For each new element, x_t, a node-selection rule defining the context value, $r_t = F(x^{t-1})$ as a node in the tree. The selected context will in turn define the conditional-
probability assignment, $P(x_t|r_t)$, by (5.3) on the basis of the occurrence counts of
the node. Arithmetic coding may be performed on the basis of $P(x_t|r_t)$.

(ii) A tree-building step over the context elements defining (and limiting) the potential
contexts. The tree stores the statistics for each node. The statistics are updated in
this step.

For each new element, x_t, the path in the context tree is given by following the
branches according to the previous elements, $(x_{t-t_1}, x_{t-t_2}, \ldots)$, until a leaf node of
the tree is reached. The context-selection principle (in Step 1) is given by choosing the
thus-far, in some sense, best node of the path.

Each node of the dynamic tree maintains occurrence counts as the statistics for the
context it represents. The node counts are updated after each new element, x_t, along the
path in the context tree given by the previous elements, $(x_{t-t_1}, x_{t-t_2}, \ldots)$, until a leaf
node of the tree is reached (in Step 2). If the combined count of the current leaf node
becomes at least 2, a new set of son nodes is added, with what was the leaf node as
their father (Step 2). The occurrence counts of the new nodes are initialized to 0, i.e. the
counts defining the probability assignment reflect occurrences in the context in the past
starting from the creation of the node.

Algorithm Context is described below in more detail for a sequence of binary symbols,
$x \in \{0, 1\}$. Further, we assume for simplicity that the elements of the context tree are
given by the symbols in reversed order, i.e. for x_t the path in the tree is given by
$(x_{t-1}, x_{t-2}, \ldots)$.

For each new symbol x_t, the tree is climbed along the path given by the previous
elements until a leaf is reached. In each node in the tree the occurrence counts (n_0, n_1)
are maintained. Each time a node is passed, the value of either n_0 or n_1 is incremented,
depending on whether x_t is 0 or 1. (The symbol x_t is coded using the counts prior to
the symbol itself.) A pair of new nodes may be added with the leaf as the father as
mentioned above as part of Step 2.

For context selection defining Step 1, efficient selection may be based on father–son comparisons at nodes s along the path in the tree defined by the causal symbols $(x_{t-1}, x_{t-2}, \ldots)$ observed previously in the sequence.

Let s denote the father node and $s0$ and $s1$ the two son nodes, i.e. the concatenation of the string corresponding to s and 0 or 1 as the prior symbol. So, if s is given by x_{t-k}^{t-1}, then x_{t-k-1} is the previous element defining the branching of the path corresponding to extending the context string by one symbol.

The ideal code lengths (5.4) of the contexts given by the two son nodes, $s0$ and $s1$, are calculated on the basis of the occurrence counts (n_0, n_1) in each node. The code length at the father node is here modified ($L'(s)$) by restricting it to the occurrences also recorded at the sons and thus the counts are given by the sum of counts recorded at the sons. Now pick the father node s over the pair of nodes of the sons, if

$$L'(s) \leq L(s0) + L(s1). \tag{5.8}$$

Note that the increment of the code length is given by the expression $-\log p(x_t | F(x^{t-1}))$, (5.4). Therefore it is sufficient to increment one variable $\Delta L = L'(s) - L(s0) - L(s1)$, by incrementing $L'(s)$ and the son node which is on the current context path. The decision (5.8) may be based on ΔL, which may be stored at the father node. A simple solution is given by applying this evaluation from the root until a father node is better (5.8) than the pair of nodes of the sons. Once the node has been selected the symbol x_t is coded on the basis of the occurrence counts of the node (5.3).

The basic version of *Algorithm Context* for binary data given above is summarized below. For a given sequence of symbols, x^T, a sequence of binary context trees, $T(t)$, is defined. For each node s in the tree a pair of counts (n_0, n_1) is stored, denoted $(n_{0|s}, n_{1|s})$. The algorithm is initialized by $T(0)$ given by the root node with counts $(0, 0)$. Thereafter, for each new symbol, x_t, perform the following steps.

(1) Traverse the context tree, $T(t - 1)$, along the path given by the context string, $x_{t-1}, x_{t-2}, \ldots$, where each substring defines a node on the path. For each new node s, compare the code length of this with the pair of sons (5.8) based on the counts. If the father node, s, is better (or a leaf node is reached) then stop. The node selected defines the context, $r_t = s$. Perform arithmetic coding on the basis of $P(x_t | r_t)$, (5.2), given by the $T(t - 1)$ occurrence counts, $(n_{0|s}, n_{1|s})$, at the selected node $r_t = s$.

(2) Traverse the context tree, $T(t - 1)$, along the path given by the context string and, for each new node s, update the occurrence by incrementing $n_{i|s}$ for $x_t = i$, until a leaf is reached. If the total count $(n_0 + n_1)$ of this leaf node, s, becomes 2, add two new leaves with s as the father node. Initialize the counts by setting $n_i = 1$ for the new leaf on the path and set the other counts to 0. The resulting tree defines the next tree, $T(t)$.

The traversal in Steps 1 and 2 may be performed as one traversal. The important issue is that the context selection and the actual (arithmetic) coding are performed on the basis of the counts of $T(t - 1)$, i.e. prior to updating the statistics as given by x_t. The

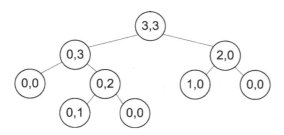

Figure 5.3. The context tree, $T(6)$, for $x_1^6 = 010101$.

comparison (5.8) in Step 1 may be performed by calculating and storing a ΔL in the father node as described above.

Example 5.6 (Algorithm Context). *Consider the sequence $x_1^7 = 0101010$. The tree is initialized by the root node only and the counts $(n_0, n_1) = (0, 0)$ at the root node. The symbols are processed one by one. After the first symbol, $x_1 = 0$, has been coded, the root count is incremented to $(1, 0)$. The second symbol, $x_2 = 1$, is also coded at the root node and thereafter the root count is updated to $(1, 1)$ and two new nodes are created. The count of the node given by the context $s = 0$ is $(n_{0|s}, n_{1|s}) = (0, 1)$. For each new symbol x_t, the tree is climbed according to the context path, $x_{t-1}, x_{t-2}, \ldots$ and the context node is selected on the basis of (5.8) as the first father node better than its sons. The value of x_t is coded using the counts of the context. The counts of the nodes along the path are updated by incrementing n_i for $x_t = i$ until the leaf has been reached. At the leaf a new pair of nodes is added if $n_0 + n_1$ becomes 2. After the first six symbols of the sequence have been processed, the context tree is as depicted in Fig. 5.3. When coding symbol $x_7 = 0$, the context $x_6, x_5 = 10$ is read by climbing the tree. For selecting the context node to be used for coding, first the root node is compared with its sons. The counts of the sons are $(0, 3)$ and $(2, 0)$. Combining these occurrences in the father node (here s is the root) gives the counts $(2, 3)$ for $L'(s)$. Selecting $\delta = 1$ in the probability estimate (5.3) allows the adaptive code length, $L(s)$, (5.4), to be expressed by (5.6): $L'(s) = \log(6!/(2!3!))$ and $L(s0) + L(s1) = \log(4!/3!) + \log(3!/2!)$. Thus $L'(s) > L(s0) + L(s1)$, and the son node representing the context $s = 1$ is chosen. Comparing this node with its sons gives $L'(s) = L(s0)$ because there have been no occurrences in the other son $(s1)$ yet, i.e. $L(s1) = 0$ and $L'(s) = L(s0) + L(s1)$, so the father node $s = 1$ is chosen as the context. The counts $(2, 0)$ of the selected context assign the estimated probability $P(x_7 = 0|r_t = s) = 3/4$ by (5.3) with $\delta = 1$.*

The version of Algorithm Context given above is simple, but in some cases it may also be too hesitant in choosing long contexts for high performance. For universal coding the full path should be evaluated, but the selection rule should still be hesitant to go for too-long contexts with little statistics. The selection rule is referred to as being *MDL (minimum description length)*-based, where "description length" refers to the adaptive code length which can describe the data considered, since the comparison between a father node and the sons, (5.8), is based on their description lengths of the symbols recorded at the son nodes.

Variations of Algorithm Context have been presented, e.g. choosing the deepest father node on the path which is better than the sons. When the full path is searched, a restriction is imposed on (the growth rate of) the context length (namely it must be smaller than $c \log t$, where c is a constant), in order to ensure universality.

A machine defined by the nodes of a complete tree is called a tree machine. Such a machine may be defined by the nodes selected by Algorithm Context. To define an actual model, the tree $\mathcal{T}(T)$ defined by the data set x^T may be pruned. Starting from the leaves, sons may be compared with the father. The sons are pruned if the father is better, but the pruning stops if the sons are better. This is repeated until no more sons are removed. The resulting tree is a tree machine.

Algorithm Context is universal in the class of finite-memory sources or tree-machine sources. It provides optimal convergence both in the mean sense and almost surely: compared with any minimal complete tree, $\mathcal{T}$, with K leaves, and probabilities $P(a|s) > 0$, the finite-memory source defined by Algorithm Context asymptotically achieves a code length for x^n within $(K(|\mathcal{A}| - 1)/2) \log n$ (plus a term that vanishes as $n \to \infty$) of the code length of the tree, $-\log P_T(X^T)$. We note that the first term $K(|\mathcal{A}| - 1)$ gives the numbers of parameters and that this number is multiplied by $\log n/2$ as in the analysis based on Stirling's approximation (5.7). This term may be seen as (a bound on) the model cost paid for learning the model parameters. The proof is based on showing that the probability of over- or under-estimating the context length tends to zero and does so fast enough. Thus asymptotically the algorithm also identifies the correct context within the class almost surely and thereby also identifies the best model within the class in this sense.

The tree machine of Algorithm Context differs from finite-state machines in that the next context is not necessarily given by the previous context and the previous symbol, x_t. This distinction is important for image data. Algorithm Context may e.g. be applied using a raster-scan traversal of the image, as was also the case for the other context-based image-coding schemes. When defining the context, a (re)ordering of the causal data is applied. As for template-based coding, Algorithm Context may readily capture 2-D dependencies in images by defining the context in two dimensions while performing sequential coding. This further generalizes to contexts in higher dimensions or the inclusion of any relevant side information.

In Chapter 2, the PRF was presented as a simple 2-D model. On applying Algorithm Context to an outcome of a PRF, the per-symbol code length will converge asymptotically to the entropy of the PRF, assuming that the three PRF context pixels are on the candidate list, which is a quite obvious choice. The algorithm will, in the almost sure sense, select the nodes defining these for each new pixel to be coded.

Actually defining 2-D models with finite contexts that yield stationary solutions is a challenge. We shall return to this issue in Chapter 7.

5.3   Dynamic context algorithms

Empirical results for Algorithm Context have indicated that, in order to achieve good results on finite data sets, e.g. image data, the selection rule should have some bias toward long contexts. This may be obtained by biasing the sons or clamping the value of the difference $\Delta = L(s) - L(s0) - L(s1)$. Algorithm Context may be applied to bi-level images, providing highly efficient compression. In order to combine speed and high compression efficiency, the father–son comparison may be extended to compare a father with the descendants in the path. This may give increased performance combining root to leaf traversal while still allowing quick access to long contexts without the operation being stopped by pixels on the path with little additional information.

Example 5.7 (Full-path context coding of binary images – FPAC).   *A variation of Algorithm Context for binary (image) data may be defined by allowing comparisons of the father not only with the son on the path, but with all descendants on the path. We refer to this scheme as the full-path Algorithm Context (FPAC). The trick is to view, say, the son not on the path as the complement of the son on the path w.r.t. the father. Thereafter this may be generalized so that the father node (s) is competing with the combination of a descendant node (sd) and the complement (s\bar{d}) of this descendant node defined by subtracting the occurrences of the descendant node (sd) from those of the father node (s). This comparison may be expressed by $L(s) - L(sd) - L(s\bar{d})$. The algorithm proceeds along the path from the root until a father, s, is better than the sons. Thereafter the descendants are compared with this ancestor node until a better node is found. This is repeated until a leaf is reached. Thus each node in the full context path is evaluated once either in competition with its father or in competition with an older ancestor. A maximum length on the context path is imposed for practical reasons. Another deviation from the basic version of Algorithm Context is that the counts in a new pair of nodes are initialized by (approximately) half the counts of the father node. The compression performance may be improved by using a greedy search based on template-based coding of the given image to determine the ordering of context pixels. In this case the (relative) positions of the context pixels are coded in the header. If the ordering of context pixels is given beforehand, the algorithm amounts to one pass as Algorithm Context. If a search for good context pixels is conducted, it is a multi-pass algorithm.*

Example 5.8 (Extending context coding of binary images – free tree).   *Another variation to Algorithm Context for binary images has its point of departure in the ordering of context pixels. Experiments have demonstrated that compression may benefit from image-dependent ordering of context pixels. This may be extended by performing the search for each node in the tree, i.e. let the context pixel defining a node depend on the path so far rather than just on the depth in the tree of the node. This is referred to as free-tree coding. In this scheme both the context tree and the context pixel at each node of* *the tree are coded in the header; only the occurrence counts in the leaves are determined adaptively. The compression requires many operations, but the decompression is actually fairly fast.*

For coding text files, and files in general, another variation of the context approach has been developed. In this case the alphabet is generally larger than just binary, and for most cases has a size of 256, reflecting an initial byte-oriented representation of the file. The scheme could also be considered a hybrid of Algorithm Context and LZW coding.

Example 5.9 (Prediction by partial matching (PPM)). *The PPM scheme essentially builds a context tree as in Algorithm Context. In the basic version, the (maximum) depth of the tree is predefined. For each new symbol the scheme looks for the longest match of the current context in the tree. If the symbol has appeared in this context, it is coded on the basis of occurrence counts; if not, an escape character is coded and the length of the match is reduced by one symbol. This is repeated until the symbol has been coded. The root, i.e. the empty string, holds all symbols of the alphabet. The approach introduces escape characters for non-binary alphabets. The escape character raises the issue of assigning a probability to the set of symbols not seen so far (in the given context). The PPM scheme is simpler and faster than Algorithm Context in the sense that context selection is just based on matching. The tree growth is controlled by a maximum depth, thus deviating from the idea of universal coding. Many variations on the basic scheme have been developed. PPM generally provides better compression than do the LZ-based schemes, but the latter are widely used due to their advantage in speed.*

Allowing the context pixels to be located in 2-D space facilitates coding of images and pixel-based graphics in the same way as for other context-based coding. For pixel-based graphics as maps, very efficient coding may be obtained.

Example 5.10 (Coding of maps). *To compare the performance of context-based coding methods with that of the graphics coders (GIF and PNG, Example 5.5) based on LZ coding regarding the results of applying some of the techniques presented to pixel-based graphics, a map of the central part of Copenhagen is shown (see Fig. 5.4 and Table 5.3). The map is 723×546 pixels and it has 12 different colors. The context-based coders are bi-level JBIG template-based coding (Example 5.2) applied to the four bit-planes (gray-level coding is applied prior to forming the bit-planes), and 2-D PPM (Example 5.9) with an optimal choice of context pixels selected in two dimensions. For comparison a state-of-the-art lossless image coder (JPEG-LS) designed for natural images rather than graphics and based on predictive coding is also included for comparison, showing that designing the coding scheme with reference to the characteristics of the (image) data is important.*

Table 5.3. Lossless code lengths (bytes) for the street map in Fig. 5.4

GIF	PNG	JPEG-LS	Template	2-D PPM
			JBIG, Gray	(Optimal order)
49 248	43 075	66 778	33 298	18 991

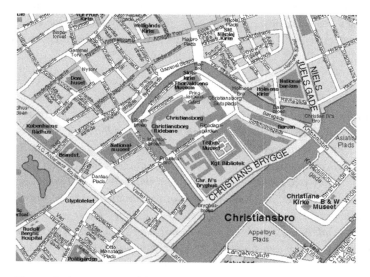

Figure 5.4. A street map of the central part of Copenhagen (©2009, Folia a/s).

It is seen that the 2-D context-based methods are significantly better than the LZ-based GIF and PNG. This is because the 2-D contexts are better at capturing the 2-D structure.

5.4   Code lengths and estimation of entropy

Estimating the entropy of a given data set, $x = x^T$, e.g. an image, is relevant in many ways, one objective being to determine bounds for lossless compression. Here entropy estimation is treated from this point of view.

Assume that the entropy, $H(X)$, is well defined, but that the model (class) is not known. For the adaptive code length, L, we have

$$E[L(x)] > H(X),$$

where $E[\ ]$ denotes the expectation. The adaptive code length will on average provide an upper bound on the entropy. This is easily shown as follows. If we assume that the conditional probabilities, $p(x_n|x^{n-1})$, are known and L' is the code length obtained by summing $-\log p(x_n|x^{n-1})$, then the average is given by the entropy, i.e. $E[L'] = H(X)$.

Having to estimate the conditional probabilities will on average lead to a larger code length, i.e. $E[L(x)] > E[L'(x)] = H(X)$.

In general, for real data, we do not know the entropy, the model order, and the model class. Thus we may see the adaptive code length rather as an upper bound on the entropy (in an average sense) than the other way around. To proceed, we may define a model class and model order k and define the entropy for the probabilities of the given class. For a stochastic variable with output x, we introduce the entropy of the model, $H_k(X) \geq H(X)$. $H_k(X) > E[I_k(x)]$, where I_k is the empirical self-entropy for that model order, i.e. using count ratios directly. Let $L_k(x)$ be the adaptive code length for this order-k model. Comparing $L_k(x)$ and $I_k(x)$ for given k bounds $H_k(X)$ from above and below in an averages sense. So the difference $L_k(x) - I_k(x)$ gives an indication of how accurate they are as estimates for a given data set. On comparing $L_k(x)$ and $I_k(x)$ for increasing model order k for a given data set, the adaptive code length will decrease to some value of k and then increase due to the effect of context dilution, whereas the self-entropy will decrease and may in the extreme case converge to 0, when each instance is mapped onto a unique context. In this respect I_k will eventually (for large enough k) provide a lower bound (on average) on the entropy. The problem is that we do not have much evidence as to when.

The template-based coding provides one model class with the order k being the number of pixels. For a given data set the adaptive code length L_k will reach a minimum for some value of k. The value of I_k will be strictly smaller than L_k and it will keep on decreasing in a (weakly) monotonic fashion.

5.5 Notes

Variations of Algorithm Context have been presented, e.g. [2], [3], [4], [5]. The context-selection rule used in the version of Algorithm Context presented here was introduced in [4]. It is simple, but not strictly universal. For sources allowing a reasonable ordering of the elements in the context tree this should not be a major issue from a coding point of view. In [3], the deepest node on the path which is better than the sons is selected. In [5] the deepest father node which is better than the son on the path is selected. In both of these solutions where the full path is searched, a restriction is imposed on (the growth rate of) the context length (namely it must be smaller than $c \log t$, where c is a constant), in order to ensure universality. This was thoroughly treated for finite alphabets. In [5], the authors go one step further than universal coding, by introducing a universal finite-memory source. It is based on their version of Algorithm Context, using tree growth and context selection, but, to complete the source, the smallest supertree, i.e. a tree containing the set of selected nodes, is chosen to define the source. The variation of Algorithm Context for binary data presented in Example 5.7 was introduced in [6], where also other context-based methods such as the Free tree (Example 5.8) were considered.

The minimum-description-length principle briefly mentioned roughly states that the model (within a class) achieving the shortest code length is also the best model. An introduction to this rich principle is given in [2]. The important characteristic is that a model cost is defined, either explicitly by a two-part code or implicitly by the adaptive code length as part of the code length (exceeding the entropy). The minimum description length balances the size of the model, which may be inferred from a given data set, providing a solution to the problem of overlearning. In this perspective, universal coding may also be viewed as unsupervised learning.

Models for two dimensions are further considered in Chapter 7, including the difficult relation of a 2-D context definition and stationarity of the model. While the 2-D context is an elegant solution to capture 2-D structures for sequential coding, the traversal of the image data is somewhat artificial. The use of hidden states combined with contexts is one attempt to remedy this, which is briefly touched upon in Chapter 7.

For practical applications aiming at a specific domain, e.g. compression of bi-level images, which is fairly well delimited, the simpler context-based techniques may be preferred over universal coding as in Algorithm Context. One example is the bi-level image-coding standard JBIG [1], presented in Example 5.2. The model order may be selected in the form of a 10-pixel template, and up to a 16-pixel template in JBIG2 on the basis of prior knowledge of the application. These techniques are also treated in more detail in the next chapter.

Exercises

5.1 (a) Derive a simple approximation for $\log n!$ in terms of bounds from above and below.

(b) Compare the results of this approximation for some values of n with the results of Stirling's approximation as given and the exact values.

5.2 Generate the dictionary and the sequence of index and character pairs $\langle i, C(x) \rangle$ which result from the parsing of the string *abracadabra abracadabra* using LZ78.

5.3 Implement Algorithm Context on the binary image below. (Define the surrounding values as being 0.) Order the context pixels by distance to the current pixel. (You may measure the distance to a point offset $(0.1, 0.2)$ from the current pixel to resolve ambiguities of distances.)

$$
\begin{array}{l}
000000 \\
010111 \\
010010 \\
010010 \\
010010 \\
000000
\end{array}
$$

5.4 Implement Algorithm Context for binary (image) data. (For simplicity you may impose a maximum length of the context.) Test it on a binary image generated by a PRF.

5.5 Calculate the ideal adaptive code length for the binary PRF. Compare your answer with the entropy of the PRF. What would you expect the redundancy to be?

5.6 Express two estimates with expected values that are an upper bound and a lower bound on the PRF entropy. Calculate the two estimated values for the PRF image.

References

[1] JBIG, *Progressive Bi-level Image Compression*, ISO/IEC International Standard 11544 (1993).

[2] J. Rissanen, *Information Science and Statistics* (New York: Springer, 2007).

[3] J. Rissanen, "A universal data compression system," *IEEE Trans. Inform. Theory*, **29** (1983), 656–664.

[4] J. Rissanen, "Complexity of strings in the class of Markov sources," *IEEE Trans. Inform. Theory*, **32** (1986), 526–532.

[5] M. J. Weinberger, J. Rissanen, and M. Feder, "A universal finite memory source," *IEEE Trans. Inform. Theory*, **41** (1995), 643–652.

[6] B. Martins and S. Forchhammer, "Tree coding of bilevel images," *IEEE Trans. Image Processing*, **7** (1998), 517–528.

6 Information in two-dimensional media

6.1 Introduction

In the previous chapters we have presented the basic concepts of information theory, source coding, and channel coding. In Chapters 1–3 we have followed traditional information-theory terminology in distinguishing between sources, which produce information, and channels, which are used for transmission (or storage) of information. In many current forms of communication information passes through multiple steps of processing and assembly into composite structures. Since in such cases it can be difficult to make a distinction between sources and channels, we use the neutral term *information medium* to refer to structures, whether physical or conceptual, that are used for storing and delivering information. In short form the terms *medium* and *media* are used. The diverse forms of electronic media may serve as examples of the composite objects we have in mind and the range of meanings of the term. As a specific case one can think of a two-dimensional (2-D) barcode printed on an advertising display so that it can be read by a cell-phone camera and used as a way of accessing the website for the business.

In the case of highly structured composite objects we shall make no attempt to directly apply concepts like entropy or capacity. Instead we limit our applications of information-theory tools to more well-defined components of such objects in digital form. The present chapter discusses how 2-D media can be described in the light of these concepts, and how the various tools can be used in such applications. The focus in this chapter is on 2-D discrete representations as in raster graphics.

6.2 Images and documents such as two-dimensional media

The 2-D media that are primarily of interest here are generated from a collection of data with the explicit purpose of making these data available to the receiver. Thus the notion of a quantity of information is relevant, and there is a potential for using it as a measure of the efficiency of the way in which data are stored or transmitted.

Some of the situations we refer to involve storing information on devices that are inherently 2-D. This includes paper, magnetic surfaces, and various optical devices. Coding for (2-D) storage applications is treated in more detail in Chapter 7. However, many applications are related to "pages" displayed on various electronic devices or printed material such as documents, maps, drawings, and newspapers (Fig. 6.1).

(a)   (b)

Figure 6.1. Bi-level image data representing the mask of an image object (a) and a layer of a street map (b) ([3] ©2006 IEEE).

Much of the 2-D content that is stored or transmitted is created by recording a three-dimensional (3-D) scene with a camera or a graphics representation thereof, thus obtaining a projection onto two dimensions, and more or less directly storing or reproducing the result. Many of the coding techniques we have described can be applied in such a setting, but it is extremely difficult to give even an approximate model of the signal.

For such a "natural" image it is not possible to give a precise definition of an information content, and such a concept would probably also be of limited relevance. However, various estimates clearly indicate that the possible (lossy) compression of an image is limited by practical and computational considerations rather than by any fundamental limit. A similar situation occurs in the coding of speech, where practical compression still leaves a transmission rate that is about two orders of magnitude greater than the estimated information rate.

6.2.1   Mixed media and document layout

Images are one component of mixed media such as newspaper pages, which combine segments of different types such as text, graphics, and images into a composite structure. Thus a description of the layout including identification of the type of data within each segment is a first step in the description. This is the function of so-called mark-up languages, which may be best known to non-professionals as the formats for describing internet pages, e.g. HTML (HyperText Markup Language).

Actually the need for representing the same information on pages of different sizes and styles dictates that the mark-up language should specify the types of the various components and their relation (sequence of text segments, relation of figures to points in the text, etc.), whereas the actual typefaces and physical placements are chosen separately.

The layout may require a significant amount of data, and the result must satisfy several constraints that have the nature of puzzles, namely covering the plane with given pieces, avoiding overlap, etc. The nature of such problems is that they cannot be described by simple languages, and usually the entropy would not be computable.

The purpose of the layout is not just to fill the page. The presentation conveys information in itself. In addition, the organization is meant to allow the reader to locate information of interest without searching through the entire text. Actually, one of the essential aspects of 2-D presentations is that the receiver is free to access the information in a sequence that is not specified by the transmitter, and that the receiver has the ability to access only those segments that are of interest. This function is served in a serial file by a table of content with related start addresses, but in the 2-D format it is often implicit in the layout. Ideally the coding should support this property allowing the receiver to search the medium by content and to decode selected segments with a moderate overhead.

6.2.2   Compound documents

Once the components and layout have been decided on, a compound document may be represented by a multi-layer and multi-level representation. This approach will also include the representation of documents by segmented regions. The composition of a document may be defined on the basis of a basic three-layer model having a foreground, a background, and a mask for blending them. The three-layer operations may be repeated to define a mapping from a multi-layer representation to one compound document. The layers may have different resolutions and they may have definitions at different resolutions. Here we describe a sequential method of blending at one resolution that is based on an ordering of layers. All *image layers*, $f_k(i, j)$, have, if necessary, been converted to the resolution of the compound image, which provides a representation ready for reproduction on a screen or in print. Initially the image is defined by a background image, $x_0 = f_0(i, j)$. In each step, the next (foreground) layer and the corresponding mask are used to define a new compound image in the sequence by blending the previous image with the next image layer, x_k,

$$x_k(i, j) = a_k(i, j)f_k(i, j) + (1 - a_k(i, j))x_{k-1}(i, j), \quad k > 0. \tag{6.1}$$

The blending function a_k may be a binary mask and represent a *shape*, or a_k may be non-binary to allow a smooth blending, in which case the mask may also be referred to as an *alpha-plane*. The compound image may also be written as

$$x_k(i, j) = a_k(i, j)f_k(i, j) + \sum_{l=0}^{k-1} a_l(i, j)f_l(i, j) \prod_{m=l+1}^{k} (1 - a_m(i, j)),$$

where $a_0(i, j) = 1$. This also gives an explicit expression of the weight for a given pixel of a given layer, $f_l(i, j)$. The layout approach with segments may be represented using binary masks. Extension to multi-resolution objects by defining the blending function (6.1) and image layers of the objects at different resolutions is straightforward. Coding of multi-resolution objects is briefly treated later.

6.3   Media components

When 2-D media are used for presenting information, there are complex processing steps involving some form of coding. In the following sections we describe some types of media and give an overview of the coding methods.

6.3.1   Printed text

When a 2-D page is used for presenting text, a standard text file clearly represents the bulk of the information involved. If the text is written in a natural language, it is at best a complicated matter to estimate the entropy; indeed, it may be questioned whether there is any such well-defined quantity. However, if we are dealing with a simpler and more well-defined problem, a program written in a standard programming language, tables of financial data or similar news, we can use finite-state sources as models and obtain an approximate value of the entropy as well as potential source-coding methods. Universal coding methods, in the form of the two Lempel–Ziv schemes introduced in Chapter 5, are the basis of the compression schemes usually applied to text files and files in general on computer systems.

Even in this relatively simple case, the conversion of a text file to a 2-D format is not a simple matter. Many of the choices and the corresponding problems will be familiar to the reader who has just a minimum of experience with text processing: a page layout has to be chosen, a typeface selected, line spacing and type sizes for headings and various types of text specified, etc.

These choices are, of course, represented as codes by the text-processing system, and the result may be obtained in the form of various standard formats for printed documents. Thus a formal language is used to present these choices, and again universal compression methods may be applied to obtain a more compact representation.

Thus it is at least conceptually clear that a page of printed text represents an amount of information, which, in addition to the text itself, includes some layout information. If a document is presented as binary raster graphics, e.g. prepared for printing, the process could in principle be reversed. The binary raster graphics may be represented in compressed formats as PDF or fax formats, which are widely used for transmission. Thus we can measure the efficiency of the compression method employed in such a transmission system by comparing it with the original document file in compressed form.

6.3.2   Maps and line drawings

Documents like maps, technical diagrams and drawings, and similar graphical presentations of data are an increasingly important way of communicating information. The explosive increase in applications involving integration of maps and information retrieval, from navigation systems to finding services via mobile phones, illustrates the importance and complexity of this field.

The information contained in such documents again contains a considerable amount of text and similar symbolic data, as discussed in the previous section. However, the symbolic information is related to points in the underlying metric grid, and the connection is made in various ways by placing the text close to relevant signatures, by use of arrows, or by other means. Different fonts, colors, and styles of text are also used extensively to indicate associations.

The metric information can roughly be described in graph terms as a number of special points, nodes with given coordinates, and lines connecting these points. Again the lines

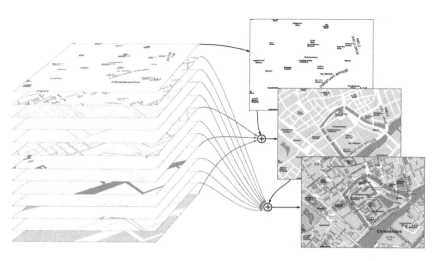

Figure 6.2. The binary layers of a city map of Copenhagen are shown together with three blended composite versions representing a progression ([4], ©2002 IEEE).

can be described in terms of their color, width, and smoothness (straight lines, parts of circles, fitted curves, etc.). Again, simple versions of this process will be familiar from the drawing facilities in text-processing systems.

Maps and other graphics are often generated in layers f_k, where the layers may represent different categories. In maps there will be roads, buildings, water, etc. The layers may be stored in a data base in various representations, e.g. in symbolic form. The data are blended (6.1) and presented in a composite form as one map, which may be printed or (at a particular time) displayed on a monitor. The map may often be the response to a query meaning that the user has expressed an interest in certain information. When the layers represent content they will naturally support content progression. We consider maps as an example of layered graphics. Figure 6.2 depicts a city map, which is composed of binary layers. Furthermore, a progression in three steps over the (content) layers yielding increasing detail is shown. The data originate from a data base where positions are given by real numbers, but the figure depicts the layers after they have been converted into a raster-graphics representation.

6.3.2.1   Layered raster-graphic maps

Consider a raster-graphics representation in which each pixel $y(i, j)$ takes on one of $N + 1$ values, $y \in \{0, \ldots, N\}$, representing a color. We shall also refer to this as the composite image and apply a raster-scan index for coding, i.e. the pixels are indexed y_t and the whole image having T pixels as y^T. The image is derived as a special case of the layered representation (6.1). Let $\mathbf{x}_t$ denote a binary vector $\mathbf{x}_t = x_t(0), \ldots, x_t(i), \ldots, x_t(N)$, $x_t(i) \in \{0, 1\}$. Each layer is assigned a color specified by the index. All pixels are assumed to be defined in at least one layer and layer 0 may simply be a background layer or a layer in its own right. A composition rule (6.1) will specify a many-to-one mapping from $\mathbf{x}^T$ to y^T. For simplicity we consider a pixel-by-pixel mapping from $\mathbf{x}_t$ to y_t and furthermore, as an example, select a mapping by priority, i.e. the layer index also reflects the order of priority expressed by

$$y_t = l_0, \quad l_0 = \max_l\{l | x_t(l) \neq 0\}, \quad 0 \leq l \leq N. \tag{6.2}$$

Thus the masks, $a_k(i, j)$, in (6.1) are binary and the layers are opaque, therefore the layer with highest index will determine the color and we may well take the layers in reverse order as is done on Fig. 6.2.

6.3.2.2 Models of discrete line graphics

Straight lines are a major component of maps, engineering drawings, etc. Consider a simple grid model of line drawings given by straight lines having two endpoints. The lines will be connected by having common endpoints. In a continuous-domain representation the endpoints may be defined by coordinates (x, y) in the plane. These may have infinite precision and thus not constitute a finite representation. Once they have been quantized to finite precision, issues determining the geometric relations may arise, e.g. whether two lines are perpendicular. The quantization may be represented by a grid of points.

The amount of information needed to specify n points and j connecting lines in an M by N grid is

$$n \log(MN) - \log n! + \log \binom{n^2}{j} - \log j!,$$

since the order of the points and lines is not important. Simply listing the coordinates and the endpoints of the lines would give a redundancy of $\log n! + \log j!$. Clearly such a simple model defined by sets of connected discrete points can easily be extended to allow a finite set of curves, several line formats, etc. The amount of information in a graphic document typically increases only slowly with the resolution. If geometric constraints are imposed (such as not allowing the lines to intersect), it may be difficult or impossible to calculate the entropy exactly.

Returning to digital lines, the issue becomes more intricate when the whole line, rather than just the endpoints, is quantized onto the grid to yield a discrete straight line. The relation between infinite-precision straight lines and the quantized versions is described below. This also describes the conversion of a vector representation into a raster-image representation.

Without loss of generality we consider a line in the first octant,

$$y(x) = \alpha x + e, \quad 0 \leq \alpha < 1,$$

where α is the slope of the line. A digital representation is given by introducing a grid with grid points given by the integer-valued points, $x = i, y = j$. For simplicity of notation, the quantization is given by truncating the value of y for each integer value of x, $x_i = i$, to yield y_i,

$$y_i = \lfloor \alpha x_i + e \rfloor, \quad x_i = i, \; \forall i \in \{0, 1, \ldots, l\}, \tag{6.3}$$

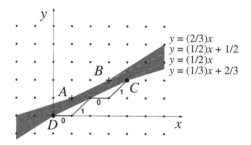

Figure 6.3. A discrete straight-line segment (DSLS) and its bounding lines ([3] ©2006 IEEE).

where l is the length of the *discrete straight-line segment* (DSLS). The $l + 1$ grid points (x_i, y_i) may be connected by l chain elements. The chain element d_i is labelled 0 if $y_i = y_{i-1}$ and 1 if $y_i = y_{i-1} + 1$. The chain elements are referred to as being 8-connected because diagonal steps are part of the set (Fig. 6.3).

It may be observed that, for a slope $0 \leq \alpha \leq 1/2$, the diagonal ($d_i = 1$) chain elements will occur singly, separated by one or more horizontal ($d_i = 0$) chain elements and the other way around for $1/2 \leq \alpha < 1$. Furthermore, the runs of the non-single element will assume only one of two neighboring values (apart from at the ends of the line segment). If we consider a discrete straight line and want to predict the next element, the observation may be used to predict the value of the next chain element of a DSLS. Actually, the longer the line segment we use, the better the prediction which is possible. For a fixed-length memory, the result is given by the preimage set defined on the parameters α and e (6.3). The preimage set of a given DSLS in the (α, e) plane is the set of continuous straight lines in the plane that are digitized to the given DSLS (6.3). It may be shown that there exist three or four lines called bounding lines (Fig. 6.3) that define the preimage set. The four lines are given by the lines with maximum and minimum slope and maximum and minimum y value at the center of the DSLS (two of these four lines may coincide, in which case there are only three lines.) The limiting points are located at the intersections of the bounding lines (A, B, C, and D in Fig. 6.3).

Each bounding line maps to a point in the (α, e) preimage plane and each limiting point maps to a line in the (x, y) plane. Thus the preimage sets are described by quadrangles or triangles in the (α, e) plane (Fig. 6.4).

We may now characterize and count the number of DSLSs. Consider a given DSLS of length l. Let p and q be relatively prime numbers such that p/q ($\approx \alpha$) specifies the slope of the center bounding line(s). It may be noted that $q \leq l$. There are $q \leq l$ different DSLSs for a line with rational slope p/q given by translating one center bounding line as far as the preimage set allows (i.e. onto the other bounding line if there are two) and noting the q instances where one or more grid points are passed. In other cases there are $l + 1$ different DSLSs, again given by the line passing each of the $l + 1$ grid points. Let $r \leq q$ index the line. Thus the DSLS is fully characterized by the quadruple (l, p, q, r). For lines in the first octant we have $l \geq q, q \geq p, l + 1 \geq r$, so the number of different DSLSs of length l is upper bounded by l^4. (Actually the number is on the

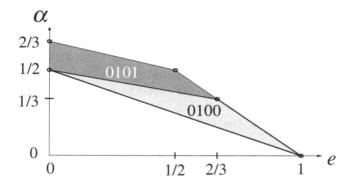

Figure 6.4. DSLS preimages in parameter space ([3] ©2006 IEEE).

order of l^3/π^2). On considering the combinatorial entropy, we see that the (per-symbol) entropy of a DSLS converges to 0 as $l \to \infty$. So, if we let the length of the DSLS and the memory used for coding increase infinitely, then the (per-symbol) entropy converges asymptotically to 0, assuming that it is indeed a straight line. Let us now consider a sequential coding of one new chain element at a time. When we extend the DSLS by a new chain element there may be one of two choices (0 or 1 in the first octant), but often the preimage set will allow only one value. When two values are possible, the next choice will bisect the preimage set (Fig. 6.4) and yield a larger value of q and a smaller preimage set. Assuming a uniform probability density in the (α, e) plane and assigning conditional probabilities for the next chain element, $p(d_{l+1}|d^l)$, on the basis of the relative areas, the sequential coding will not exceed that of the combinatorial entropy (on average.) The development of the p/q approximation of the slope α is intimately connected to a continued-fraction approximation of α.

Consider the question about the geometric property, namely whether two DSLSs are perpendicular. The question of whether there exist two preimage lines that are perpendicular is easily answered on the basis of the preimage sets: add (or subtract) $\pi/2$ to (or from) the α values of one of the preimage sets, whereafter the intersection of the preimage sets shall be non-empty, if two such preimage lines do exist. Even when this is the case, at the same time there are infinitely many pairs of preimage lines within this intersection that will not be perpendicular. For the DSLS, the preimage sets define a region of tolerance. Asymptotically ($l \to \infty$) the preimage sets will converge to one α value, and in this setting the answer may be provided.

Other geometric structures will display the same properties, but will be even more complex in terms of expressing the relations.

6.3.3 Natural images

Natural images or similar graphics are not only important as separate applications, but also are components in documents. Since they are produced as projections, or are made to be interpreted as projections, they are typically characterized by contours that are the outlines of 3-D objects against the background. Within these contours the properties are often stationary (smooth or characterized by uniform texture). These properties lead to problems closely related to the coding of the metric information described in the previous section. So a rough model can be obtained by describing an image by the edges representing discontinuities at the contour or object boundaries and locally stationary patches besides the edges.

Coding of image data is treated later in this chapter. In such media as newspapers and magazines, and when images are reproduced using a printer, the images are halftoned as a precoding step matching the essentially binary reproduction techniques.

6.3.3.1　A model of halftone images

A halftone image is a bi-level representation of a natural (gray-level) image. The offset-printing technique used in newspapers and magazines is essentially binary (ink or no ink). For color images the three basic colors of the tri-chromatic subtractive color system (cyan, magenta, and yellow) are used. Often the three colors are combined with a black for the practical reason of reducing the amount of ink. The resulting color representation is known as CMYK.

Since each component is treated separately, we just consider the black-and-white representation of the gray scale. The local density gives an impression of the gray-scale image in that area. Thus we may consider the (analog) process of halftoning as a pulse–area modulation scheme. In digital form an increased resolution is used to represent the gray-scale value.

There are several halftoning methods. We introduce the threshold halftoning method as a simple model capturing the basics. In discrete form, given an input gray-scale image, $f(i, j)$, the binary halftone image, $b(i, j)$, is given by

$$b(i, j) = \begin{cases} 0 & \text{if } f(i, j) > t(i, j), \\ 1 & \text{if } f(i, j) \le t(i, j), \end{cases} \tag{6.4}$$

where $t(i, j)$ is a 2-D periodic function. In color reproduction the threshold function, and thereby the halftone grid, has a distinct angle for each color in order to reduce the beating of frequencies known as moiré.

Using (6.4), the binary value at each position represents the information concerning whether the image value is above or below the threshold function. The discrete functions in threshold halftoning (6.4) could be replaced by continuous functions.

For traditional halftone images as used in newspapers, the threshold function may be described as a 2-D periodic and continuous function, e.g. a normalized 2-D cosine function. With centering of the threshold function at the maximum value, the cross sections will decrease monotonically away from the centering. For slowly varying images this gives clustered dots of black or white color. Dispersed dots or dithering may be generated by (6.4) using a discontinuous function. Dithering of the least significant bits of a gray-scale image may also be applied to increase the perceived amplitude precision, again at the expense of precision in spatial position.

For a different approach, error diffusion may be used. At each position the error is calculated and diffused to the non-causal neighboring values.

6.3.3.2 Error diffusion

It is often useful to model a stochastic signal in one dimension as the output of a linear filter that is excited by a random process with a white spectrum (mutually independent inputs in discrete time). Compression of speech can be based on a modified representation whereby the input to the filter is a sequence of binary pulses. This type of process is not so attractive from a formal point of view, but its use is motivated by the mechanism by which speech is generated in the vocal tract. This model also immediately suggests an encoding composed of a description of the filter (assumed constant over a significant period of time) and a binary excitation sequence. For a filter of finite impulse response the mechanism has a finite number of states, and in principle we can find the input sequence that gives the best match to a given waveform.

In the reproduction of a gray-scale image we can assume that the eye detects a weighted average over a small segment of the field when the details are too fine to be resolved. Thus a desired level of gray can be produced by a pattern of black and white with the desired average density. We consider a process of generating such a pattern on a grid with a separation between pixels that is about an order of magnitude below the resolution of the eye.

Unfortunately, there is no practical way of finding the 2-D sequence which gives the best fit to a given image, even for very simple 2-D weighting functions (filters). As an approximation, the commonly used approach is a heuristic method known as *error diffusion*. Let the input be image samples $x(i, j)$ with values between 0 (white) and 1 (black), and let the output be the binary printed pattern, $y(i, j)$. If the input with value $x > 1/2$ is converted into a black output, $y = 1$, there is an error, $e = x - y$, for which we will compensate in the encoding of the following pixels (taken in the causal order). This is accomplished by splitting e into a sum of terms

$$e = \sum_{i',j'} e(i', j') = \sum_{i',j'} w(i', j')e \tag{6.5}$$

with positive weights, $w(i', j')$, where (i', j') denotes the shift of that part of the error. These error terms are then added to the next input values before they are encoded. Let $d(i, j)$ denote the resulting error diffused into the pixel at $y(i, j)$; the error is $e(i, j) = x(i, j) + d(i, j) - y(i, j)$, where $y = 1$ if $x + d > 1$, $d = e(-i, -j)$. Intuitively, the error-diffusion process will cause the binary output pattern to have approximately the desired average density.

Actually, an analysis of this nonlinear 2-D system is a highly challenging task, and we shall just make some comments on a few simple situations. Assume that all weights are rational and that the input is a constant rational value $x = p/q$ between 0 and 1. It would be desirable that the output is a periodic 2-D function with the required average value and the smallest possible period, u (at least the denominator of x). If such a pattern can be found, the error terms would also be periodic, and they would have to satisfy a system of linear equations. For simplicity, but without loss of generality, we state them for a rectangular 2-D period $u = r \times s$ with sides r and s, indexed 0 to $r - 1$ and 0 to $s - 1$:

$$e(i, j) = p/q + d(i, j) - y(i, j), \quad 0 \leq i < r, 0 \leq j < s,$$

$$d(i, j) = \sum_{i', j'} w(i', j') e(i - i', j - j'), \quad 0 \leq i < r, 0 \leq j < s,$$

where the indices $(i - i', j - j')$ are calculated mod(r, s), i.e. the errors being diffused out at one side of the period are identical to those being diffused in on the other side.

If the numerical error is at most 1/2 in all u cases, such a pattern would actually continue once the encoder was started in the correct state. On the other hand, adding the equations above clearly gives zero, and thus the solution is not unique. Whether the system would converge to a particular pattern is not readily decided. For practical applications several weight sequences have been suggested, and their properties have been studied experimentally.

It has been observed that error diffusion tends to give a good reproduction of edges, compared with other methods of halftoning. This property may be explained by noting that, following a dark area, the diffused error terms tend to be negative, and an adjacent light area will be encoded as white close to the boundary.

We may note that the gray-scale representation has higher resolution than the resolution of the period of the threshold function, but lower resolution than that of the binary image. High-contrast parts such as edges will have high spatial representation but uncertain amplitude values, whereas low-frequency content will have a good amplitude representation.

6.4   Context-based coding of two-dimensional media

Two-dimensional media such as graphics may often be generated in a symbolic representation, but later the data are displayed on a pixel-based device. Digital means of acquisition such as scanning printed material will also produce a pixel-based representation, and, for transmission of the data, a pixel-based representation is widely used. Here we consider coding the 2-D data in a pixel-based format or for pixel-based reproduction, i.e. as raster graphics. The relation to the underlying structure is also considered.

The starting point is the context-adaptive coding introduced in Chapter 5. On the basis of a local causal neighborhood, r_t, of the current pixel, x_t, the context is defined and an adaptive conditional-probability estimate, $p(x_t | r_t)$, is obtained. Arithmetic coding is performed on the basis of this estimate. Once the arithmetic coding is in place, the focus is on the model defining the context function $r_t = F(x^{t-1})$ and (to a lesser degree) the adaptive probability estimate based on previous occurrences in each context. Actually, the definition of contexts is very general and the context may be defined on the basis of higher-dimensional data sets or given side information. The critical issue is that the index t reflects a sequential processing of the data defining the causal data available (at the decoder) when (de)coding an element, x_t.

In Chapter 5, a simple adaptive probability assignment leading to the adaptive code length was introduced. The process of learning the parameters incurs a cost on each context, so for a given data set or application a challenge is to strike a balance between having a good distinction in the model that is based on many contexts and the model cost. Algorithm Context provided an elegant universal solution by allowing the set of contexts (in the context tree) used for coding to grow. Practical applications are characterized by finite data sets and often domain knowledge such as the approximate order of magnitude of the size of the data sets and some idea of the inherent structures. Thus, for most applications, the context function $F(x^{t-1})$ is fixed on the basis of domain knowledge and only the probability assignment is adaptive.

For bi-level images the context function may efficiently be defined by a subset of nearby pixels, e.g. a template with 10–16 pixels. For non-binary images the number of contexts grows very rapidly and it is generally necessary to define the context in two steps by first selecting a set of pixels and thereafter performing an additional quantization.

If the data are defined by layers, the template may readily be defined across layers by selecting pixels from previous layers. Examples are graphics composed of layers such as maps or multi-resolution representations going from coarse to fine resolution. This approach may serve both progressive and interactive applications.

6.4.1 Lossless context-based coding

Defining the context function by a template of causal pixels is a simple yet effective way to define the contexts for coding bi-level images. For images over larger alphabets a template is a good starting point. The templates may be selected to capture the structure of the image data. Consider a binary source with template pixels X_{t-t_i}, each having the value c_i, that is the value of the template pixel having a fixed position relative to the current pixel, X_t. The context may simply be indexed by reading the binary number given by the value of the template pixels, $\sum_i c_i 2^i$. For each context the pair of counts $(n(0), n(1))$ of occurrences is maintained. The probabilities are estimated on the basis of these and thereafter are used to drive the arithmetic coder. We note that by using the counts obtained on the causal part of the image the decoder may mimic the encoder, enabling the decoder and encoder to remain synchronized.

Example 6.1 (Bi-level image coding using adaptive template pixels – JBIG). *Templates are efficient for coding bi-level images. Often the nearby causal pixels are used to define the context function. For some types of image data, selecting other data-dependent positions of the template pixels may capture the structure better. Coding of halftone images with a periodic structure is an example, both for dither images and for clustered-dot halftones. One or more of the template pixels may be placed one period or some number of periods away. In this way the threshold function $t(i, j)$ will have the same value at the current position and at the positions of these template pixels. The ISO standard JBIG2 allows up to 16 pixels with up to four adaptive template pixels. For JBIG the* *rightmost pixel of the 10-pixel template is adaptive. For bi-level images templates of up to 20–30 pixels may give a coding gain, but clearly the complexity increases and the return is diminishing.*

The template captures local structures or, using adaptive template pixels, more remote information, but over a fixed limited set of pixels. As mentioned in the introduction, straight lines (and straight boundaries) are important elements of graphic material such as line drawings, maps, etc. Using a fixed template provides only a small window through which the line or boundary is seen only locally. As an example we may consider the three-line template of JBIG (Chapter 5) and the set of digital straight lines forming a boundary dividing the template pixels into black and white pixels. Given three elements of a DSLS, all combinations of two chain elements are possible, due to the quantization. More chain elements are required for the DSLS structure when it is desired to enforce deterministic elements of a DSLS.

The free-tree coding introduced in Chapter 5 may provide higher performance on such material by starting with the nearby pixels and selecting the next context pixels according to their values, thus potentially tracking the straight line. For perfect straight lines the free tree should ideally select the next non-deterministic pixel.

6.4.2   Context quantization

For non-binary images the number of contexts becomes very large even for templates of modest size. One approach that can be employed to overcome this is to define the context function, $F(x^{t-1})$, as a composite function. First a template selects the context pixels and thereafter the value is mapped onto a smaller set of values. This is referred to as *context quantization*. Often the context quantization may be combined with a decomposition of the value, x_t, to be coded. First a context quantizer applicable to graphics material will be considered. Thereafter the issue of optimizing a context quantizer will be treated.

6.4.2.1   Relative pixel patterns

In graphics the current pixel will often take on the same color as one of the (causal) neighboring pixels. As indicated by the straight-line analysis, the pattern further away may also convey additional information about the current pixel. A simple context-quantization scheme for graphics is to map the pixels within the template onto an index of the pattern. Consider the template pixels, c_i, in a given order. Label the first pixel value, c_0, by A. The second pixel value, c_1, is also labelled A if the values are identical, i.e. $c_1 = c_0$. If they are different, c_1 is labelled B. The template pixels are considered one at a time and a template pixel is given a new label if it is different from all the previous template pixels. The resulting quantized template represents a pattern, but not the actual value.

The number of different contexts, $C(\tau)$, for a template with τ pixels may be determined recursively. Let $C_j(\tau)$ denote the number of the $C(\tau)$ contexts which have exactly j different pixel values and let $|\mathcal{A}|$ denote the alphabet size. On adding one template

Table 6.1. The number of distinct contexts, $C(\tau)$, as a function of the template size, τ, using relative patterns.

τ	1	2	3	4	5	6	7	8	9	10
$C(\tau)$	1	2	5	15	52	203	877	4140	21 147	115 975

pixel, the number of contexts $C_j(\tau + 1)$ is given by adding one of the j colors already appearing in any of the $C_j(\tau)$ contexts or adding a new color to any of the $C_{j-1}(\tau)$ contexts. For given $|\mathcal{A}|$, the recursion is

$$C_j(\tau + 1) = C_{j-1}(\tau) + jC_j(\tau), \quad 1 < j \le |\mathcal{A}|, |\mathcal{A}| \le \tau < \infty,$$

which is initialized by $C_1(\tau)$, $1 \le \tau$ and $C_j(\tau)$, $j > \tau \vee j > \mathcal{A}$. Calculating $C_j(\tau)$ and summing up gives the total number of contexts for a template of size τ,

$$C(\tau) = \sum_{j=1}^{\tau} C_j(\tau). \tag{6.6}$$

The number of contexts $C(\tau)$ for templates of size up to $\tau = 10$ and $|\mathcal{A}| > \tau$ is given in Table 6.1. Even a template size of $\tau = 10$ is manageable irrespective of the alphabet size. The required memory may likewise be determined as $P(\tau) = \sum_j jC_j(\tau)$ since each pattern with j distinct values has j free parameters. A counter of an escape symbol for the case in which x_t is different from the j values present may well be used, leading to $j + 1$ counters in a practical adaptive scheme. The analysis may also be used for fast indexing of the contexts.

Example 6.2 (Coding of graphics by relative pixel patterns). *Efficient coding of graphics may be achieved by defining the context by quantization by relative pixel patterns. In the first step it is coded if the pixel has the same value as one of the context pixels and if not an escape character is coded. In the case of an escape, the pixel value may be coded using the same RPP context. As a faster alternative all the escape pixels may be collected in one string, which is coded using a standard Lempel–Ziv-based file-compression method (See Chapter 5.)*

Example 6.3 (Piece-wise-constant coding of graphics). *A related and very efficient coding of graphics may be obtained by first decomposing the coding of the pixel value into four questions coding whether the current pixel has the same value as one of the four causal neighbors one by one. First the binary question of having the same value as the previous pixel in the same row is coded. An edge-map is used to define the context quantization. Given two four-neighboring pixels, the binary edge values are defined according to whether the two pixels are identical or not. In piece-wise-constant (PWC) coding the causal edges of the four neighboring pixels are used to define the context. If the current pixel is different from the previous pixel, the pixel above is considered using* *the same context. If none of the causal four neighbors are identical to the current pixel, a guess list is consulted and, if necessary, the actual pixel value is finally coded. This example presents the basic principle of a very efficient scheme.*

6.4.2.2   Optimal context quantization

Finding *the* best model is an undecidable problem also for source coding. Instead we may pose the question of what the best context quantizer for a given data set is. Consider a given set of contexts. These may be defined by a subset of the causal data, $C_t = X_{t-t_1} \ldots X_{t-t_k}$. The k context pixels drawn from the alphabet $\mathcal{A}$ lead to $\mathcal{A}^k$ contexts. This may be too many in terms of learning the statistics or from an implementation point of view due to memory requirements.

First we consider the issue of mapping the contexts C_t onto one of M indices using a context quantizer $Q(C)$. Assume that the conditional probabilities $P(X|C)$ are given. Clearly $H(X|Q(C)) \geq H(X|C)$ by virtue of the convexity of H. So in this setting the problem is that of how to minimize the increase in conditional entropy, i.e. minimize $H(X|Q(C))$ or, equivalently, minimize the divergence of $D(P(X|C)||P(X|Q(C)))$ weighted by the context occurrence. The optimal solution may in general be quite complex and need not necessarily be connected in the context space described by k pixels. The solution is much simpler when defined in terms of the conditional probabilities, $P(X|C)$. Viewed in this way, the optimal solution is given by convex sets in the probability simplex of X. To simplify further, we consider a binary random variable, Y, and consider the terms $H(Y|C)$ and $H(Y|Q(C))$. The optimal solution is given by intervals of the variable $Z = P(Y = 1|C)$. The entropy of the minimum-conditional-entropy context quantizer (MCECQ) with M values, $P(Y = 1|Q(C))$, is given by

$$H(Y|Q(C)) = \sum_1^M P(Z \in [q_{m-1}, q_m)) H(Y|Z \in [q_{m-1}, q_m)), \qquad (6.7)$$

where the set of thresholds $\{q_m\}$ divides the unit interval into M contiguous intervals (with $q_0 = 0$ and $q_M = 1$). Given $P(Y = 1|C)$, the optimal MCECQ may be found by searching over the set of thresholds, q_m, defining the probability intervals. This problem may efficiently be solved by dynamic programming, providing the exact solution.

In practice we do not have $P(Y = 1|C)$ available, but for a given data set we may estimate the conditional probabilities. Besides providing an efficient solution for binary data ($X = Y$), the solution (6.7) may also be used in combination with a binary decomposition of the variable X to be coded.

For a given data set we may also evaluate the quantizer in terms of the (ideal) adaptive code length. With this approach the optimal solution also provides a value for M. Consider the ideal adaptive code length

$$L(x^T|C) = \sum_1^T -\log \hat{p}(x_t|c_t), \qquad (6.8)$$

where $\hat{p}(x_t|c_t)$ is the adaptive probability estimate of x_t in the context in which it appears. As noted in Chapter 5, the adaptive code length may be expressed in terms of the occurrence counts. Let L_m denote the code length in the quantized context indexed m and defined by the contexts c mapping onto m, i.e. $Q(c) = m$. The minimum-code-length context quantizer, $Q(c)$, is given by

$$\min_{M,Q(c)} \sum_{m=1}^{M} L_m = \min_{M,Q(c)} \sum_{m=1}^{M} L(x^T|Q(c)). \tag{6.9}$$

Again we consider the binary case and follow the approach of MCECQ by defining the context quantizer by contiguous intervals of the conditional probabilities, $\hat{p}(Y = 1|c)$, obtained on the whole data set. Again the set of interval ends $\{q_m\}$ defines the quantizer, and they may efficiently be determined by dynamic programming. If the estimates are given by $\hat{p}(Y = 1|c) = n_1/(n_0 + n_1)$ then we conjecture that the fast solution is indeed the solution defined by (6.9) in the binary case. Since the code length includes a cost for learning the parameters in each quantized context, the solution provides a finite M balancing the parameter cost and the accuracy of the conditional probabilities.

The expression of the minimum code length (6.9) does not include the cost of coding the context quantizer, $Q(c)$. Therefore the value (6.9) gives a lower bound on the adaptive code length achievable for x^T given the raw context set C, had we had the foresight to select the optimal context quantizer. The approach may be used for training and evaluating context quantizers. It may also be used adaptively for an image sequence. In this case it would be possible to optimize the context quantizer on the previous image for use when coding the current image.

6.4.3   Lossy context-based coding

Lossy coding of 2-D data may be designed for specific media applications. This will later be briefly treated using model-based coding for specific 2-D material and for natural images. Generic lossy coding of 2-D data poses at least two major challenges. The first is that of how to define a good distortion measure, and even for simple distortion measures such as the Hamming-distance rate-distortion analysis remains intractable. In Chapter 3, the rate-distortion function for an i.i.d. binary source with probability $P(0) = p$ was derived as $R(D) = H(p) - H(D)$ for the Hamming distance. Taking the obvious next step of considering correlated sources renders the problem intractable.

A simple and general, albeit not $R(D)$-optimal, approach is to change a (small) number of pixel values prior to the context-based coding. This approach goes hand in hand with the Hamming distance. The number of altered pixels gives the Hamming distance. This has the practical advantage that the encoder and decoder do not have to be redesigned, but the coding disadvantage that the coding does not exclude the so-eliminated configurations (though they may be assigned relatively small probabilities).

The question for this approach is that of how to control which pixel values to alter. For context-based adaptive coding it is possible to express the effect on the *operational* rate-distortion performance of changing the value of one given pixel, i.e. the effect on bit rate and distortion for a given coding scheme. A scheme for *bit-flipping* may be devised on this basis.

The context-adaptive code length, $L(x^T)$, (6.8), may be expressed on the basis of the occurrence counts in each context. The order of occurrence does not influence the code length. Changing the value of one pixel, X_t, which has a context r_t will have the "direct" effect of changing the occurrence counts in context r_t. Without loss of generality we may express the "direct" effect, ΔL_0, in the binary case in which $x_t = u$ is flipped to the complementary value $\bar{u}$ in the context $r_t = c$. The count $n_u(c)$ will be decremented and $n_{\bar{u}}(c)$ incremented, leading to the difference in the adaptive code length (6.8),

$$\Delta L_0 = \log\left(\frac{n_u(c) - 1 + \delta}{n_{\bar{u}}(c) - 1 + \delta}\right). \tag{6.10}$$

Besides the direct effect, the flipping will also have an influence on all the pixels for which x_t is part of the context. For template-based coding the set of pixels for which x_t is part of the template is given by mirroring the template around the current pixel, x_t.

Let u' denote the value of the mirror pixel of template pixel number q, and let v_q and w_q denote the original and new context values, respectively, of u' when flipping X_t. Thus one occurrence of u' is moved from context v_q to context w_q and the code length changes by

$$\Delta L_q = \log\left(\frac{n'_u(v_q) - 1 + \delta}{n_0(v_q) + n_1(v_q) - 1 + 2\delta} - \frac{n'_u(w_q) + \delta}{n_0(w_q) + n_1(w_q) + 2\delta}\right). \tag{6.11}$$

Now combining the direct effect (6.10) with the indirect effect (6.11) of the k pixels whose context is affected by the flipping of x_t gives the total marginal effect on the adaptive code length (6.8) caused by flipping x_t,

$$\Delta L = \sum_{q=0}^{k} \Delta L_q. \tag{6.12}$$

Any bit-flipping scheme may use the marginal change to decide which pixels are flipped. It may be used for evaluating one pixel at a time once the statistics for the whole image have been collected. If one evaluates more than one pixel at a time the effects may interact both by virtue of one candidate being a context pixel for another and, less pronouncedly, by virtue of more than one pixel altering the statistics of the same context.

A simple greedy scheme is given by the following procedure.

(1) Collect image statistics, $n_0(c)$ and $n_1(c)$, for all contexts c.
(2) Evaluate candidates according to a threshold and possibly flip them
(3) Repeat until the desired value of L or an acceptable value of d has been obtained.

Step 1 provides the statistics for the evaluation by ΔL. The selections may be based on evaluation of the marginal effect on the (operational) rate-distortion performance $\Delta L/\Delta d$, where Δd is the marginal effect on distortion. For the Hamming distance, $\Delta d = 1$ is constant. The pixels are traversed in a given order and the pixel having a better $\Delta L/\Delta d$ is flipped. The statistics are updated after each flip. The image may be traversed multiple times in evaluating whether more pixels can be flipped. The threshold is changed in such a way as to slide down the operational rate-distortion curve $L(d)$. This may be repeated until a desired value of L or an acceptable value of d has been achieved or no more pixels will decrease the code length (or improve distortion).

The bit-flipping scheme may also be used as an unsupervised noise filtering. Consider a document that has been scanned into a binary format. Assume that the noise is uncorrelated and also uncorrelated with the image content. The context-based coding will capture structure from the image data. On running the bit-flipping scheme, the noise bits will with high probability be corrected to comply with the structure of the image.

6.5   Context-based multiple representations

In many applications involving communication of image data, it may be of interest to have multiple representations, e.g. to help cope with the large amount of data of an image data set. It may be that the communication system and the decoding time impose a delay so that the full image is not perceived as being retrieved instantaneously. In this case progressive coding may provide the user with useful data right away and thereafter the information is refined. This may involve resolution, quality, and content progression. In other cases the data set, e.g. a large map, is larger than what may be reproduced on a screen or perceived by the user. In this case multiple resolutions are commonly used, e.g. on the Internet. In other cases there may be different reproduction capabilities on the computer screen, on the screen of a mobile device, and for the printer. So both a difference of resolution and spatial subsets as regions of interest may be useful.

As mentioned, the flexibility of context-based coding and templates provides for sequential coding of 2-D and higher-dimensional data. Besides the two spatial dimensions, the templates may represent the dimension(s) of progression.

6.5.1   Refinement coding for quality progression

Lossy compression of bi-level images was introduced above. We may want to refine this to a lossless representation. Let $b(i, j)$ denote the original bi-level image and let $b_r(i, j)$ denote the reference image, here the lossy version coded first. For refinement of the pixel at (i', j') from lossy to lossless it is natural to select a template of causal pixels in $b(i, j)$ and combine it with the pixel (i', j') and surrounding pixels in $b_r(i, j)$. Once the split template (Fig. 6.5) has been defined, context-adaptive coding may be carried out. The approach may be generalized to multiple steps, whereby in each step a refined version of lower distortion is coded on the basis of the previous version as reference. This is referred to as progression by quality.

Example 6.4 (Refinement coding – JBIG2). *Refinement coding is used in the JBIG2 standard, thus providing progression by quality, including refinement to lossless. It is the choice of the encoder to make the decisions and e.g. control that the distortion is actually reduced. Also the decoder does not know per se whether the final result is lossless or not. The split template is composed of the four causal neighbors of the current image and nine pixels in the reference image, namely the coinciding pixel and its eight neighbors.*

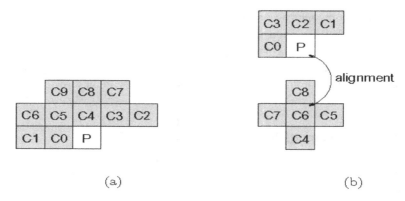

Figure 6.5. A template (a) and a split template making reference to another image (layer) ([3], ©2006 IEEE).

Refinement coding may also be performed by applying distributed source coding based on coding correlated sources (Chapter 3), yielding a quite different solution from the context-based refinement coding. The reference (or side-information) frame, X, is used when coding the current frame Y without having access to X. In image coding X may represent a lossy version. This may be coded using any compression scheme as long as the quality is above some limit. Thereafter the distributed source coding is applied to code Y given X at a rate not less than $H(Y|X)$ but having access to X at the decoder side only.

6.5.2   Multi-resolution context-based coding

Multi-resolution representation of image data, $f(i, j)$, may be obtained by subsampling the data set. A simple choice is to reduce the resolution by a factor of 2 in each dimension, giving $g(k, l) = f(i/2, j/2)$ for $i = 2k$ and $j = 2l$ even. This may be repeated a number of times to yield a multi-resolution or pyramid representation. Instead of simple subsampling, the image data may be low-pass filtered prior to each subsampling step. Multi-resolution using a subband or wavelet transform is presented briefly later. Here we consider binary images and graphics.

A multi-resolution representation may be coded starting with the low-resolution image and thereafter coding the next higher resolution conditioned on the previous resolution.

Example 6.5 (Multi-resolution representation of binary images). *For binary images containing text and line drawings, the images are sparse, being mostly white, with a relatively smaller number of black pixels conveying the information. Simple subsampling or low-pass filtering would after a few decimations most likely render the image difficult to interpret and, if low-pass filtering were applied, more or less white. For this type of material a bias toward black pixels or toward maintaining the topology of the objects of the image makes sense. A simple choice when replacing four binary pixels at one resolution by one binary pixel at a lower resolution is to select a black pixel if at least* *one of the higher-resolution pixels is black. For coding from coarse to fine resolution a split template may be used. In JBIG a split template with six causal pixels at the current pixel, $b(i, j)$, is used together with four pixels of the lower resolution including $b_r(i/2, j/2)$. There are four "phases" reflecting the relative position of $b(i, j)$ and the corresponding pixel at the lower resolution. Each phase has its own template definition with a bias in the non-causal direction. The subsampling scheme may be specified by the user, but the default scheme is a heuristic scheme designed for the sparse images of text and line drawings, as mentioned above.*

6.5.3   Context-based coding of maps and layered graphics

Layered raster graphics was introduced previously (6.1). On the other hand, any given composite image, y^T, having $N + 1$ colors may be mapped (one-to-one) onto a layered binary representation by

$$y_t = \begin{cases} x_t(j) = 1, & j = i, \\ x_t(j) = 0, & j \neq i. \end{cases}$$

The layers resulting from the mapping are referred to as split layers.

The layered approach may naturally be generalized to a hybrid representation having (coding) layers that may be non-binary. Let $\mathbf{y}_t$ represent a vector of pixels from L (coding) layers, $\mathbf{y}_t = y_t(1), \ldots, y_t(L)$, where each $y_t(i)$ takes on one of a subset of the $N + 1$ colors. The pixels of the hybrid layered image $\mathbf{y}_t$ will also map onto the pixels of a composite image.

6.5.3.1   Coding of layered graphics

Context-based coding of multi-level raster graphics was treated previously. A composite image, y^T, may also be coded by extracting the bit-planes (going from most significant bit (MSB) to least significant bit) and coding the bit-planes as binary images. The bit-planes may be coded individually, but increased performance may be achieved by placing context elements in previous bit-planes, e.g., when starting with the MSB bit-plane, combining the bits of the current pixel from more significant bit-planes with causal bits of the current bit-plane.

Here we consider coding the layered representation, $\mathbf{x}^T$. The first scheme is coding each binary layer, $x(i)^T$, as a binary image. This provides a layer-progressive coding and, if the layers represent different content, it is also *content progressive* and the flexibility is high because any order of transmitting or decoding the layers may be chosen. Higher compression may be obtained using cross-layer templates. This will obviously introduce causality among the layers and thereby constraints on the progression order. The dependency may be described by a *directed-dependency graph*, where for each layer it is noted which layers are directly used when coding it. Maps may well share information across layers and there may be layers that are obvious choices as reference layers. Allowing template pixels selected from *one* other layer, all pairwise combinations

Table 6.2. Lossless code lengths (bytes) for the composite street map

RPP (9 pixels)	PWC	2-D PPM	Bit-plane (direct)	JBIG (Gray)	Template (extracted layers)	Skip coding
21 002	20 600	18 991	33 298	30 426	27 682	20 700

may be tried out and dynamic programming over the directed-dependency graph may be used to select the best order from a compression point of view.

If we consider a layer-progressive coding of x^T for progressive display ending up with y^T and assume the priority mapping (6.2) then we may note that the color of a pixel $y_t = i$ may already be determined if the index i has a higher priority than the current layer with index j. Thus the value of $x_t(j)$, $j < i$, is void w.r.t. the color of the display pixel, y_t. Therefore we may skip the coding of this pixel. If we do so, we must ask what value we should assign the pixel when it is used as a context pixel. Besides the value of $x_t(i)$, another option is context quantization by assigning the background value N for context definition, i.e. $y_t < i \Rightarrow c(y_t) = N$. If we consider this layered approach as a coding of y^T, it involves one-by-one decomposition into the colors and context quantization both by selection of the layers used and by the value-skipping quantization above.

6.5.3.2 Results for coding of maps

The street map (Fig. 6.2) was coded both in the ordinary (composite) representation and in the layered representation (Table 6.2). The context-based techniques suited to graphics (RPP, Example 6.2, and PWC, Example 6.3) provide similar results and work almost as well as the 18 991 bytes achieved by the (more complex) variable-context-length 2-D PPM coding (Chapter 5) operating with adaptive context lengths. Template coding of the bit-planes of the pixel values (y_t) is simple but not competitive. Gray coding the pixel values prior to the coding of bit-planes does help. As mentioned above, binary layers may be extracted from the composite image and thus provide content progression. When skip coding is introduced into a template coding of the binary layers almost the same result as for PWC is achieved, while providing a layer-progressive coding at the same time.

The layered representation, providing progression by layers, was also coded. Even with standard template-based coding using a ten-pixel template (JBIG), results are almost as good as for the coding of the composite image. Template coding brings it to the same level and free-tree coding (see Chapter 5) reduces it to 17 768 bytes. Template coding with split templates with the option of selecting template pixels in one causal layer reduces it further. Finally, selecting the best coding method for each layer reduces the length to 14 878 bytes (seven of the layers were coded using the free tree and the rest using a reference layer) (Table 6.3).

It is interesting to note that, while the layered data set must have a higher entropy than the composite image expressed by $H(\mathbf{x}) \geq H(y)$, due to the mapping of $\mathbf{x}$ to y, actually the coding of the layered representation is the most efficient. This means that for this

Table 6.3. Lossless code lengths (bytes) for the layered street map

JBIG	Template	Template plus reference	Free tree	Best
22 761	20 804	16 905	17 768	14 878

domain there is a strong structure within the layers and to some extent between layers pair-wise and, furthermore, that it has been easier to capitalize on the domain knowledge in the layered coding. It also means that it should be possible to do significantly better coding of the composite image, y.

6.6 Model-based coding of two-dimensional document components

Previous sections have described how discrete representations may efficiently be coded using context-based coding. Various dependencies such as spatial, temporal, and cross-layer may all be addressed using (split) templates (Fig. 6.5). While the statistical dependencies may efficiently be captured within local neighborhoods or split neighborhoods given by relatively few elements, long-range dependencies, especially dependencies involving many elements, are more difficult to capture due to the explosion of the number of different contexts. In many cases the discrete 2-D representation is generated from a fairly compact symbolic representation. In documents, text and halftones are prominent examples, as well as the line drawings already discussed. Here we consider the so-called model-based coding, which is designed for specific sources and to some extent mimics the generation of the raster-graphic representation based on a symbolic representation. An extreme example would be to resort to the program generating the data itself. This is the case when storing and communicating a graphics data file in the internal format of the graphics program. There are two reasons why one should seek a more generic representation for communication purposes. One is the independence of proprietary formats and different versions; the other is that, for reproduction of a specific 2-D discrete instance, coding the full data set used to produce it need not be efficient, e.g. consider a single map at a given resolution for a given application versus (the part of) the data base used to generate it.

Example 6.6 (Binary documents). *A document may be described as composed of text, images that are represented as halftones, and graphics. The bi-level image-coding standard JBIG2 offers coding of symbol patterns (aimed at text) and halftone patterns, and generic coding in terms of template-based coding. The generic coding alone may represent any bi-level image data, but the model-based coding aimed at text and halftones may provide increased compression performance and features when reproducing the data. JBIG2 does not provide a designated model for line drawings. In some cases the symbol-pattern coding could be applied, but in general it is envisioned that the generic code will be used to represent these.*

6.6.1   Symbol-based coding of text

Text is an important source of 2-D data in documents in general. Some examples are business documents and newspapers and magazines as well as web-pages. While template-based coding does provide efficient coding of bi-level images, including text images, the template captures only local structure. The template is typically smaller than a character, so the longer-range structure of characters being repeated is not fully captured. Later we shall return to issues of layout, but here we just consider the basic 2-D representation of the characters in a bit-map representation.

Text may efficiently be coded using universal schemes as presented in Chapter 5. A font is used to define the graphic appearance of the characters. Using a single font, this provides an efficient representation. In the general setting of having all fonts as well as capabilities of supporting all languages, the character library may well become so large that the representation becomes inefficient (from a data-compression point of view) and also the reproduction becomes complicated. Here a symbol-based approach without explicit coding of fonts etc. is considered. It is based on a dynamic dictionary of symbols, such that only the symbols used are actually coded. The bit-map to be coded may be generated directly by a word-processing program or it may be scanned documents. The exposition follows the latter more complicated problem, which may encompass the other case as a noiseless scanning. The approach has similarities to optical character recognition (OCR), but, rather than being recognition, it is an issue of reproduction. So on the one hand it is not necessary to recognize the character (or even the alphabet or whether it is a character), on the other, for high-quality reproduction the character should be reproduced with small enough distortion that the appearance due to the font also appears to be reproduced.

6.6.1.1   Pattern matching and substitution

A basic approach to coding based on pattern matching and substitution is outlined by the following steps.

- Segment into pixel blocks.
- Search for a match.
- Code the match if it is acceptable.
- Encode the bit-map if there is no match and include it in the dictionary.
- Encode the position of the pixel block (as an offset).

The dictionary may initially be empty or primed by an application-specific set of bitmaps. The pattern matching above suffers from the same problem as OCR, namely a small risk of substitution errors for scanned material (or low compression efficiency).

A crucial issue is the distortion measure to decide whether the match is acceptable. A simple choice is the Hamming distance, which may be elaborated by combination with connectivity measures.

6.6.1.2   Soft pattern matching

To reduce or even eliminate the risk of substitution errors, the pattern matching above may be extended with a refinement coding of the symbols. The refinement coding may be template-based, using the match as reference and coding the current bit-map (or a version with a small acceptable distortion) using a split template. This provides efficient refinement coding and thus a low cost for cleaning small errors. Further, the new symbol in the dictionary may be coded conditional on the identified best match. Lossless coding is feasible, each time refining the symbol to lossless. In this case not all coded pixel blocks should be included in the dictionary.

Example 6.7 (Pattern matching in JBIG2). *The ISO JBIG2 bi-level image-coding standard supports both pattern matching and substitution (PM&S) and soft pattern matching (SPM) as well as combinations of them. In PM&S the new pixel blocks for the dictionary are not coded interleaved with the coding of matched symbols, but instead one or more dictionary segments are used and separately a data-segment block codes the index of the match and where to position the pixel blocks. In the PM&S coding Huffman coding is used for fast coding. In SPM arithmetic coding is used for high compression efficiency and for efficient refinement coding. SPM may provide lossless coding directly, or alternatively a clean-up coding using the generic refinement coding may be applied for lossless coding. Lossy coding is controlled by the encoder.*

6.6.2 Coding of halftone images

Halftone images have been described in terms of periodic halftones and error diffusion, which are not periodic. Here a model for periodic half-tones (with clustered dots) is introduced. Let (x, y) denote the real-valued image coordinates and (i, j) the integer-valued pixel positions of (x, y). The halftone grid is given by the basis vectors, $(x_r, -y_r)$ and (y_r, x_r). For a clustered-dot halftone a priority function may well be approximated on the basis of the function along the grid coordinates (i_1, i_2)

$$s(i, j) = \cos\left((i_1 + a)\frac{2\pi}{M}\right)\cos\left((i_2 + b)\frac{2\pi}{M}\right). \tag{6.13}$$

As mentioned above, the period of halftone grids is often rotated at an angle, which may be described simply by rotating $s(i, j)$. The function $s(i, j)$ is not normalized, but an implicit normalization is implied if it is used to determined the *priorities* of a given discrete sampling.

A coding scheme may simply be based on an estimate of the underlying gray-scale image at the encoder and at the decoder (re)screening defined by an input gray-scale image, the priority function, and the grid vectors as coded by the encoder. The gray-scale image may be coded by image-coding techniques. The relevant alphabet size may depend on the halftone-cell size. The priority function and the grid vectors are parameter data that do not require many bits to code in comparison with the image data.

Example 6.8 (Halftone coding in JBIG2). *Halftone coding based on descreening and (re)screening is supported by the ISO JBIG2 bi-level image-coding standard. The scheme follows the approach of the symbol-based coding designed for text (Example 6.7).*

Therefore a halftone dictionary of halftone patterns within a rectangular (bounding box) may be coded as a bit-map. As a simpler alternative, a default dictionary based on the priority function defined by the 2-D cosine function (6.13) and the grid vectors may be used. The gray-scale image is coded as bit-planes after gray coding. This will efficiently support images of any number of bits per pixel. The bit-planes may be coded using the generic template-based bi-level coding which is used for other data within the standard. Once the gray-scale image and the halftone dictionary have been decoded, the decoder can insert the halftone "symbols" defined by the gray scale at the position defined by the grid vector and a predefined order of inserting the symbols. For angled grids, i.e. the cases in which the basis vectors of the grid are not aligned with the axis of the bit-map, the (rectangular bounding boxes of the) symbols overlap. The symbol bit-maps are combined using a logical pixel-wise or function in the case of overlapping pixels. The encoder should take this process into account for efficient coding.

6.7   Image coding

Images may be coded on the basis of conditional probabilities as described previously. If we consider natural images having 8 bits per pixel, the number of contexts if one is using more than a few context pixels becomes very large. To avoid this, we may consider other means of capturing the correlation between neighboring pixels. Two techniques are widely used, namely prediction and transforming (blocks) of the images. If lossy coding (at some level of fidelity) is acceptable, decorrelating the data prior to quantization is an efficient approach. This is widely used and many good books have been written on the topic; here we give a brief introduction.

The basic approach is given by a decorrelation step, quantization for lossy coding and entropy coding using e.g. Huffman or arithmetic coding. We consider lossless coding and assume a reversible decorrelation step, which would render a statistically independent output that could provide the basis for an optimal scheme. Recognizing that this is not the case for natural-image sources, improved performance may be achieved by applying context-based coding as part of the entropy coding to capture residual correlation in the data. In this step context quantization may be applied.

A simple approach to lossless image coding is given by bit-plane coding and applying, say, a template-based coding such as JBIG to each bit-plane. This may provide acceptable results for images with fewer than 8 bits per pixel, but predictive coding will generally give better results for 8 bpp (and down to about 5 or 6 bpp).

6.7.1   Predictive image coding

Predictive image coding is based on a prediction, $\hat{x}_t = F(x^{t-1})$, for each new pixel that is derived from the causal pixels in a row-by-row scan of the image. The prediction error, $e_t = x_t - \hat{x}_t$, is coded and the pixel value is reconstructed at the decoder, $x_t = \hat{x}_t + e_t$.

A basic predictor is given by a linear predictor

$$\hat{x}_t = \sum_i a_i x_{t-t_i}, \tag{6.14}$$

where a_i are constants defining the filter coefficients of the linear filter. These coefficients may be determined as the coefficients providing the minimum mean-square error on the basis of the correlation function for the data (possibly with the constraint that the sum of coefficients is unity). Let x_W, x_N, and x_{NW} denote the causal pixels to the west, north, and northwest of the current pixel x. In the lossless version of the JPEG standard one of seven simple predictors involving x_W, x_N, and x_{NW} may be selected. The prediction error is coded using Huffmann coding or, as an option, arithmetic coding.

A drawback of the linear filter for image coding is that (natural) images are not globally stationary. A simple modification is to consider the images as being only locally stationary with edges as transitions. It is consistent with this simple model that nonlinear prediction filters may be used. A simple yet effective prediction filter is given by

$$\hat{x} = \min(x_W, x_N, x_{NW}) + \max(x_W, x_N, x_{NW}) - x_{NW}. \tag{6.15}$$

This filter switches among x_W, x_N, and $x_W + x_N - x_{NW}$, and it implicitly involves a simple edge-detection capability.

For images the distribution of the prediction errors e will center around 0 and decay to each side. This is well modelled by the two-sided geometric distribution (TSGD),

$$p(e) = C(\theta, s)\theta^{|e-\mu|},$$

where e is an integer, $\theta \in (0, 1)$, $\mu = R - s$, $s \in [0, 1)$, and $C(\theta, \mu) = (1 - \theta)/(\theta^{1-s} + \theta^s)$ is a normalization factor. For natural images, the prediction errors are typically correlated with each other and also with the local image statistics, e.g. a higher variance is expected in texture areas and at edges. Context-based techniques may be applied to capture these remaining structures. The parameter space may be reduced by using a probability distribution, such as the TSGD, requiring only two parameters (θ, μ) per context and applying context quantization.

For fast implementation, optimal prefix codes may be derived for the parameter space. This may be done on the basis of Golomb codes, which are optimal for the one-sided geometric distribution (OSGD) having a distribution of the form $(1 - \theta)\theta^x$, $0 < \theta < 1$, for integer non-negative x. Golomb codes were introduced for run-length coding. The probability distribution of run-lengths is an OSGD, assuming that the sequence was generated by a Markov chain. The Golomb code has one parameter, m, which is a positive integer and defines the Golomb code G_m of an integer $x \geq 0$. The code G_m codes x in two parts. The integer part $\lfloor x/m \rfloor$ is represented by a unary code and the remainder $x \bmod m$ is represented by a (modified) binary code using $\lfloor m \rfloor$ bits if $x < 2^{\lceil \log m \rceil} - m$ and $\lceil m \rceil$ bits otherwise. The Golomb codes are especially simple for the special case $m = 2^k$. The k least significant bits are appended to the unary representation of the higher-order $\lfloor x/2^k \rfloor$, which may easily be calculated, e.g. by using shift operations.

The positive and negative values, e, of a TSGD may simply be interleaved to a sequence $0, -1, 1, -2, 2, \ldots$, which may be described by the indices,

$$M(e) = 2|e| - u(e), \tag{6.16}$$

where $u(e) = 1$ if $e < 0$ and 0 otherwise is an indicator function for the negative values of e. After the mapping (6.16), a Golomb code may be used. For $s \leq 1/2$, $M(e)$ is a natural mapping producing a sequence of non-increasing probabilities. The combination of interleaving and a Golomb code is referred to as $G_m(M(e))$. For $s > 1/2$ the symmetry leads to the corresponding mapping, $M'(e) = M(-e - 1)$.

The optimal prefix codes for TSGD may be defined on the basis of Golomb codes as a function of the parameters (θ, μ). The constructions are simple but non-trivial. The interleaved Golomb codes $G_m(M(e))$ are optimal for certain combinations of parameters.

Let $\Gamma_k = G_{2^k}(M(e))$ denote the distribution given by combining the interleaving with a G_{2^k} Golomb code.

Example 6.9 (Lossless image coding, JPEG-LS). *Lossless image coding may be performed using the nonlinear prediction filter (6.15) followed by modelling the prediction error by the TSGD. Contexts are introduced by making the estimates of the two TSGD parameters dependent on image structures. The northeast pixel, x_{NE}, is included in the context formation, so the four causal pixels, x_W, x_N, x_{NW}, and x_{NE}, are combined and quantized to define the coding contexts. First the differences of neighbors are calculated, namely $g_1 = x_{NE} - x_N, g_2 = x_N - x_{NW}$, and $g_3 = x_{NW} - x_W$. Thereafter each difference g_i is quantized to yield q_i, and finally the context is given by the quantized triple, (q_1, q_2, q_3). For each context the integer part of the mean (or bias) value μ is subtracted to center the distribution as $\theta^{|e+s|}$, where $s \in [0, 1)$. A simple technique is applied to adaptively estimate θ in each context. For simplicity, the interleaved Golomb codes Γ_k are used, resulting in only a small sacrifice of optimality. For each context the parameter is adaptively estimated by*

$$k = \min\{k' | 2^{k'} N \geq A\},$$

where N is the number of occurrences and $A = \sum |e|$ in the given context.

Higher compression may be obtained by applying arithmetic coding. In this case A/N is quantized and used as the context, where occurrence counts are maintained to provide the estimates.

Furthermore JPEG-LS applies run mode for efficient coding of uniform areas, such as in graphics such as the street maps.

6.7.1.1   Lossy predictive image coding

By quantizing the prediction errors, higher compression will be achieved at the expense of introducing a loss. To avoid allowing the encoder and decoder to drift, the encoder reconstructs the lossy version and uses these values in the prediction (6.14) and (6.15), so the encoder and decoder remain synchronized. A simple linear quantizer is given by

$$q(e) = \text{sign}(e) \left\lfloor \frac{|e| + \Delta}{2\Delta + 1} \right\rfloor,$$

where Δ specifies the quantizer step, which is $2\Delta + 1$.

This is referred to as near-lossless coding, since it is efficient for small values of Δ, whereas, at higher distortion levels, the use of transform-based coders and coding in a frequency domain is more efficient.

6.7.2 Transform-based image coding

To decorrelate an image a transform may be applied. An ideal would be to apply the unitary transform, which has the most compact representation in terms of having the most energy concentrated in the first k coefficients. Recognizing that images are not stationary, and in order to reduce complexity, a common approach is to divide the image into blocks of $M \times N$ pixels and apply a transform to these. A good approximation to the optimal decorrelation (for a simple image model having an exponential decay of the correlation function) is provided by the discrete cosine transform (DCT). The $N \times N$ DCT is given by

$$F(k, l) = c(k)c(l) \frac{2}{N} \sum_{i=0}^{N-1} \sum_{j=0}^{N-1} f(i, j) \cos\left(\frac{\pi k(i + 1/2)}{N}\right) \cos\left(\frac{\pi l(j + 1/2)}{N}\right),$$

where $c(0) = 1/\sqrt{2}$ and $c(k) = 1$ for $k \neq 0$. It may be noted that the transform is separable and may be performed by applying the 1-D DCT first in one and then in the other direction. The 1-D DCT may be expressed by first mirroring the signal with N samples and thus defining an even function with $2N$ samples and thereafter applying the discrete Fourier transform. The separable form of the transform may be written in matrix form as $\mathbf{F} = \mathbf{CXC'}$, where $\mathbf{X}$ is the image block and $'$ denotes the transpose, and the elements, $c(k, i)$, of $\mathbf{C}$ are given by

$$c(k, i) = c(k) \sqrt{\frac{2}{N}} \cos\left(\frac{\pi k(i + 1/2)}{N}\right).$$

Since the transform is orthonormal, the inverse is given by the transpose and therefore the 2-D inverse is given by $\mathbf{X} = \mathbf{C'FC}$.

Example 6.10 (DCT-based image coding, JPEG). *The JPEG standard applies the DCT to 8 by 8 blocks. The coefficients, $F(k, l)$, are quantized using a scalar frequency-dependent quantizer $Q(k, l)$ for lossy coding. Thereafter the coefficients $F(0, 0)$ are DPCM coded, coding the difference relative to the $F(0, 0)$ of the block to the right. A zig-zag scan is applied to the 63 other coefficients within an 8 by 8 matrix. The positions of nonzero coefficients in the scan are run-length coded. The actual coding is based on tables that may be perceived as Huffman tables. Arithmetic coding may be used as an alternative.*

To improve coding performance for the 2-D DCT, the image may be tiled into blocks of different size ($M \times N$), to capture the structures of the image better, and the DCT applied to these image blocks. As an alternative approach to capture the residual redundancy after application of the DCT, the coefficients may be organized into subbands, i.e. all coefficients $C(k', l')$ for a given frequency, (k', l'), may be collected into a smaller image. Thereafter the coefficients of each subband may be represented by bit-planes and finally each bit-plane may be coded using a binary context-based coder.

6.7.3 Subband- and wavelet-based image coding

The block-based DCT coding provides a location of the signal in space (in terms of the blocks) and in frequency (by the DCT). Subband or wavelet analysis provides another approach to space–frequency representations, avoiding the block structure and providing a multi-resolution representation. The discrete subband/wavelet transforms we consider are based on a 1-D filter pair followed by downsampling to half the sample rate for each filter output. The 1-D filter pair considered is given by the finite-impulse-response filters

$$y_0(n) = \sum_i h_0(i)x(n-i),$$

$$y_1(n) = \sum_i h_1(i)x(n-i),$$
(6.17)

where $h_0(n)$ and $h_1(n)$ are the coefficients of a low-pass and a high-pass filter, respectively. These filters are referred to as analysis filters. The outputs y_0 and y_1 are downsampled by deleting every second sample. With an appropriate choice of filters the transform is orthogonal. A separable 2-D subband/wavelet is defined by using the 1-D filters and downsampling first in one and then in the other direction. For a dyadic discrete wavelet transform the decomposition is applied again to the low-pass–low-pass (LL) subband obtained by low-pass filtering and downsampling, first in one and then in the other direction. Repeating the decompostion of the resulting LL subband yields a multi-resolution representation. At the decoder side the signal is upsampled in one direction by first inserting samples of value 0 at positions where the signal has been deleted. This gives the values

$$y_0'(n) = ((1 - (-1)^n)/2)y_0(n),$$

$$y_1'(n) = ((1 - (-1)^n)/2)y_1(n).$$
(6.18)

Finally the signal is reconstructed using a pair of synthesis filters,

$$\hat{x}(n) = \sum_i (g_0(i)y_0'(n-i) + g_1(i)y_1'(n-i)).$$
(6.19)

The application of 1-D synthesis filters is repeated to match the operations on the encoder side. With an appropriate filter design $\hat{x}(n) = x(n)$ will provide a reconstruction of the original values at the encoder side (6.17). The Haar wavelet is a simple example and it may be described by $h_0(0) = h_0(1) = 2^{-1/2}$ and $h_1(1) = -h_1(0) = 2^{-1/2}$, and $h_0(n) = h_1(n) = 0$ for all other values of n. Owing to the downsampling this may also be seen as a 2×2 transform. The short filter length implies the shortcoming that the

Table 6.4. Coefficients for the four-tab Daubechies low-pass filter, h_0

$h_0(0)$	$h_0(1)$	$h_0(2)$	$h_0(3)$
0.482 963	0.836 516	0.224 144	−0.129 410

Table 6.5. Coefficients for the low-pass filters of the 9–7 wavelet transform

$h'_0(0)$	$h'_0(\pm1)$	$h'_0(\pm2)$	$h'_0(\pm3)$	$h'_0(\pm4)$
0.602 949	0.266 864	−0.078 223	−0.016 864	0.026 749
$g'_0(0)$	$g'_0(\pm1)$	$g'_0(\pm2)$	$g'_0(\pm3)$	
0.557 544	0.295 636	−0.028 772	−0.045 636	

separation into high and low frequencies is not very effective. A class of filters is obtained from one low-pass filter with impulse response $h_0(i)$, $0 \le i \le N$. The high-pass filter is given by $h_1(N - i) = (-1)^i h_0(i)$ and the synthesis filters are given by $g_0(i) = h_0(N - i)$ and $g_1(i) = (-1)^i h_0(i)$. The filter h_0 must satisfy that perfect reconstruction (6.19) is obtained by $\hat{x}(n + N) = x(n)$. This is obtained for the Haar filter given above and, as another example, by the four-tab Daubechies filter h_0, which is given in Table 6.4. On the basis of this filter (6.17) defines an orthogonal discrete wavelet transform. (The additional properties defining a wavelet transform are outside the scope of this text.) The discrete orthogonal discrete wavelets are restricted to filters having an even length.

Unfortunately the Haar wavelet is the only (real-valued, compactly supported) orthogonal wavelet having a symmetric (or linear-phase) low-pass filter. In order to achieve odd-length symmetric filters a bi-orthogonal transform is used. We restrict the transforms to the case in which the synthesis filters are defined by the analysis filters and we require perfect reconstruction, $\hat{x}(n) = x(n)$.

In lossy wavelet-based image compression, a dyadic discrete wavelet transform is first applied (including downsampling), then the coefficients of the resulting transform are quantized, and finally the quantized coefficients are coded.

Example 6.11 (Wavelet-based image coding). *A bi-orthogonal wavelet filter is preferred in image coding in order to allow symmetric filter responses of odd length. A popular bi-orthogonal filter pair (6.17) is the 9–7 wavelet filter given by the symmetric low-pass analysis ($h'_0(i) = 2^{-1/2} h_0(i)$) and synthesis ($g'_0(i) = 2^{-1/2} g_0(i)$) filters in Table 6.5.*

The high-pass filters are given by $h_1(i) = (-1)^i g_0(i + 1)$ and $g_1(i) = (-1)^i h_0 (-i + 1)$. The filter pair has the property $\sum_i h_0(i) g_0(i + 2k) = \delta_k, 0$, i.e. the sum is 1 iff $k = 0$. The 9–7 filter is the default filter for lossy coding in JPEG2000. The wavelet coefficients are represented by bit-planes and the bit-planes are coded using context-adaptive arithmetic coding. The number of bit-planes coded will implicitly define the quantization. Starting from the most significant bit-plane, a progression over quality is given. The details are omitted from this brief description.

Example 6.12 (Distributed source coding for progressive coding of images). *Assume that we have a lossy low-quality representation of an image at the decoder side. The bit-planes of subbands of a DCT or a wavelet representation of an image, X, may then be coded using distributed source coding without the encoder having access to the low-quality representation, as side information, Y. For each bit-plane of coefficients the encoder sends syndrome bits. The decoder then corrects the lossy version of the images on the basis of the syndrome bits.*

6.8   Coding of compound documents

Many documents are a mix of text, images, line drawings, graphics, etc. Compression schemes for the diverse components have been presented. Having the individual components or layers, f_k, it may be natural to select a specific image-coding scheme for each component or layer.

Having the mixed components of a document, these may be coded and decoded independently for maximum flexibility and random access. As illustrated, utilizing cross-layer dependencies may provide more efficient coding. In this case the components or layers may be coded and decoded in a given sequential order, or for random access a dependency graph and identifiers of the components may be established to represent which components need to be available to decode a given part of the information of the document. Furthermore, rules for composing the document in terms of the components and their relative positioning are required, e.g. it can be done by blending of layers (6.1) or composition of segmented regions.

Besides blending, the composition may also add a residual image, $r(i, j)$,

$$x_k(i, j) = a_k(i, j)f_k(i, j) + (1 - a_k(i, j))x_{k-1}(i, j) + r_k(i, j), k > 0.$$

A refinement coding, e.g. using a split template (Fig. 6.5) may be seen as a way to code residual image information.

6.8.1   Coding document segments

A natural approach to coding compound documents is to segment the document into regions and, for each region, select an appropriate coding, e.g. one of the schemes introduced previously. For each region a position is also coded, often by coding the upper-left-hand corner and size of a bounding box. In terms of blending (6.1) the masks a_k are binary and these masks segment the document into non-overlapping regions for pure region classification.

Example 6.13 (Access, layout, and file structures). *JBIG2 is organized in terms of data segments. In random-access mode the segment headers are placed at the beginning of the file. This enables the decoder to construct (or extract information from) the full* *dependency graph. JBIG2 may combine symbol-region coding, halftone-region coding, and generic-region coding. A generic refinement coding may also be applied afterwards to improve the quality, e.g. to lossless coding. Furthermore, JBIG2 data segments can be embedded in other file formats such as PDF, TIFF, and MRC. JBIG2 is used as a "compression filter" in the PDF format, which has become an open standard, ISO 32000-1:2008.*

6.8.2   Coding mixed raster content

Mixed raster content may often be efficiently represented on the basis of multiple image layers and masks and selecting coding algorithms for the specific layers. It was shown that schemes for coding natural images did not perform well on graphic material such as street maps, the reason being that in such graphics the information lies in the edges and there is no natural relation of pixel values of different colors.

Example 6.14 (Mixed raster content).   *Mixed-raster-content representation based on the principle above has been defined in ITU-T T.44. It supports the blending model (6.1) with one background and N pairs of masks and image layers. For binary data context-based arithmetic coding such as JBIG2 or Huffman-based MMR coding may be chosen. For image data the simpler DCT-based JPEG coding or the more complex wavelet-based JPEG2000 coding may be chosen. A simple example is a document with an image and text where part of the text is written on top of the image. The image part is the background (coded with JPEG2000). The text is represented by the mask (using JBIG2), and the foreground codes the color(s) of the text. In this way the sharp edges of the text part may be removed from the image-coding part.*

The coding of mixed raster contents on the basis of layers and using different coders poses the question of how to optimize the coding. This was already discussed and illustrated in the case of coding layered maps with a focus on compression efficiency. Another aspect is fast (random) access. If we allow lossy image coding, as is often the case for natural images, improving the (operational) rate-distortion performance may be the objective of the optimization.

6.9   Notes

This chapter focussed on coding documents and graphic materials in a raster-graphics format. A very brief introduction to the large field of image coding was given. There are many good books on image coding covering DCT- and wavelet-based coding among other techniques. In [1] image and video coding is introduced. Wavelet-based coding, in particular JPEG2000, is thoroughly treated in [2]. In contemporary wavelet image coding, including JPEG2000, context-based arithmetic coding plays an important role and the context quantization is applied to achieve efficient block-based embedded coding. Context-based coding of bit-planes also plays an important role in current work on distributed image and video coding. As noted in the introduction, edges are an important characteristic of images and the actual coding of edges is a weakness in block-based DCT coding. Wavelets are more efficient at coding edges, since the high-frequency basis functions are more localized, but the coding is still not optimal. Finding a good unified approach to coding a mix of image and graphics is an interesting challenge; so far, using multiple coders provides the best results.

MPEG video coding is based on JPEG coding. The temporal correlation is captured by first, for each block, finding the best matching block in a previous image. After subtracting the best match in a so-called motion-compensation step, the residual is coded by applying a block-based DCT similar to that of JPEG. In object-based video coding, object information is coded in the form of shape and possibly alpha-planes. Split templates are used for coding video shape information in MPEG-4, and context-based coding may also be used for efficient coding of video alpha-plane information [3].

Exercises

6.1 Consider a three-line template used for binary images as in JBIG. Calculate the number of distinct (JBIG) contexts that can occur when a digital straight line forms the boundary of a black and white object passing by the pixel to be coded.

6.2 Calculate an upper bound for how many bits are required asymptotically to code an infinite digital straight line using an LZ78 coding.

6.3 Calculate the probabilities within the quantization intervals for a uniform quantizer, assuming a Laplacian distribution of the variable prior to quantization.

6.4 Consider a Pickard random field as introduced in Chapter 2. Implement a simple subsampling by selecting $g = f(2i, 2j)$ and thereafter perform a template-based coding of the lower-resolution image g. Thereafter, using g, perform a refinement coding to represent f. The coding may be evaluated in terms of entropy or a code length. Compare the performance of this "multi-resolution" coding with the entropy of the Pickard random field.

References

[1] J. M. Woods, *Multidimensional Signal, Image, and Video Processing and Coding* (Boston: Kluwer, 2002).

[2] D. S. Taubman and M. W. Marcellin, *JPEG2000, Image Compression Fundamentals, Standards and Practice* (Amsterdam: Academic Press, 2006).

[3] S. M. Aghito and S. Forchhammer, "Context-based coding of bilevel images enhanced by digital straight line analysis," *IEEE Trans. Image Proc.*, **15** (2006), 2120–2130.

[4] S. Forchhammer and O. Riis Jensen, "Content layer progressive coding of digital maps," *IEEE Trans. Image Proc.*, **11** (20062), 1349–1356.

7 Constrained two-dimensional fields for storage

7.1 Introduction

This chapter describes fields of discrete symbols over finite alphabets $\mathcal{A}$ (with $|\mathcal{A}|$ symbols). Such fields can serve as models of graphics media or of other forms of two-dimensional (2-D) storage medium. By addressing the storage medium as a surface on which an image array may be written, the density may be increased. Besides optical storage, new storage techniques based on holography and nano-technology have been demonstrated. Two-dimensional barcodes have also been designed for coding small messages, but increasing the capacity compared with conventional barcodes, which carry information in one dimension only. The Datamatrix is an example of such a code. We focus on 2-D constrained coding for storage applications, but the models presented are more general and they may be used for other applications as well.

7.2 Two-dimensional fields with a finite constraint

The symbols are placed in a regular grid (commonly referred to as a lattice) and indexed $a(i, j)$, where i, j are integers. The first index indicates rows and the symbols are spaced at equal intervals within the row. We usually consider rectangular grids on which the symbols are aligned in columns indicated by the second index. However, the symbols in row $i + 1$ can be shifted a fraction of a symbol interval relative to row i. In particular, a shift of half a symbol interval gives a hexagonal grid. The mutual dependency of symbols is often described in terms of neighbors. In a lattice a neighborhood refers to a set of points with a fixed set of distances from a given central point. As a common case the neighbors of a particular symbol in a rectangular grid are the four symbols above, below, and to the right and left. In other common cases the four symbols on diagonals are included as neighbors. On a hexagonal grid, there are six closest neighbors.

Unless specified otherwise, the symbols are assumed to be on a rectangular grid. It is usually convenient (and consistent with applications) to consider a finite area of M by N symbols. In that case the indices are chosen as $1, \ldots, M$ and $1, \ldots, N$ in the same way as in a matrix.

The structure of the field can be described in various ways, but we consider only the most elementary form of structure, that of a finite constraint, as introduced in Chapter 2.

For every case of a certain configuration of symbols, usually an L by L square, the total set of $|\mathcal{A}|^{L \times L}$ combinations is divided into an admissible and a non-admissible set. The non-admissible set is also referred to as the forbidden set. Thus only outcomes in which every configuration belongs to the admissible set are possible. The set of admissible configurations on an M by N rectangle is denoted $E(M, N)$. The set of neighbors of such a constraint can include any symbol within $L - 1$ rows and columns of the given symbol.

In one dimension a finite constraint defines which combinations of symbols can occur within a window of a certain size. We have discussed earlier how such finite-state sequences can be counted, and how they are related to finite-state Markov chains.

It turns out that in two dimensions even this simple definition allows too many complicated situations. It may be very difficult, or even impossible, to decide whether there are large outcomes that satisfy the constraint (thus many common puzzles have the form of covering part of the plane with pieces according to certain rules).

We are interested in the case in which the constraints are easily satisfied, and many outcomes are possible. In fact, one of the fundamental problems will be to find the number of solutions for a large rectangular section of the plane.

In Chapter 2 the hard-square constraint and some other simple examples of 2-D constrained fields were given. We conclude this section by listing these and several other finite constraints that define fields with useful properties. More detail will be provided in the examples throughout the chapter.

- Run-length-limited constraints in two dimensions.
- Fields with controlled shapes, tiling with specified figures, lines, etc.
- Crossword puzzles.
- Binary fields with specified minimum distances between 1s.
- Binary fields with restricted densities of 1s within L by L squares (e.g. random halftone images).
- Fields with restricted composition of several colors.
- Fields with restricted transitions between colors, no isolated bits, and smooth contours.

Example 7.1 (The hard-square constraint revisited). *The hard-square constraint introduced in Chapter 2 may also be expressed by requiring a minimum distance of 2 by the 1-norm between any two elements with the value 1, i.e. $x_{ij} = x_{kl} = 1 \Rightarrow |i - k| + |j - l| \geq 2$. It may even be expressed by requiring a minimum Euclidean distance of $\sqrt{2}$ between any two elements with the value 1. As pointed out in Chapter 2, it may also be described as a run-length-limited constraint, 2-D RLL $(1, \infty)$, with a maximum run length of one black pixel in any row or column.*

Example 7.2 (Minimum-distance constraints). *A generalization of the hard-square constraint is minimum-distance codes, in which all 1s must be surrounded by 0s such that*

Figure 7.1. An SRLL(2, 3) configuration ([4], ©1999 IEEE).

there is a minimum distance between any two 1s,

$$x_{ij} = x_{kl} = 1 \Rightarrow \|i - k, j - l\| \geq d.$$

Besides the distance, the norm by which it is measured is the parameters specifying the constraint. The Euclidean distance (2-norm) is obvious, but also the 1-norm and inf-norm may be considered. The shape of the surrounding set of 0s is (discrete) circular, diamond, or square, respectively.

A further generalization is to require a pattern of 0s surrounding each 1. This will not be treated explicitly, but the minimum-distance constraint will serve as an example of this class. The RLL constraint may also be imposed on both colors of a binary field.

Example 7.3 (Symmetric RLL (SRLL) constraints). *For a binary field the run-length constraints can also be symmetric in the sense that they are imposed on both colors. Thus a symmetric RLL (d, k) imposes the (d, k) constraint on both colors in both rows and columns. It is denoted a 2-D SRLL (d, k) constraint (see Fig. 7.1).*

7.3 Counting two-dimensional patterns with constraints

When the field of symbols is defined on an M by N rectangle, the admissible set refers to L by L squares that are wholly contained within the large square. Thus the symbols along the boundary can be chosen somewhat more freely. Since it is a finite problem, we can count the number of different admissible outcomes, $F(M, N)$, for each (M, N) rectangle. For simplicity, we consider squares, i.e. $M = N$, with $F(N)$ admissible configurations. As for the one-dimensional (1-D) case, we then make the following definition.

Definition. The *combinatorial entropy*, C, of the 2-D constraint is given by

$$C = \lim_{N \to \infty} \log F(N)/N^2, \tag{7.1}$$

assuming that the limit exists. C is also referred to as the capacity of the constraint.

The combinatorial entropy is calculated using binary logarithms and expressed in bits/symbol. To obtain a positive entropy, the number of outcomes must increase exponentially with the area of the square.

For any N, $\log F(N)/N^2$ is an upper bound on the combinatorial entropy. To prove this claim, we cover an arbitrarily large jN by jN square with independently chosen N by N squares. Thus the number of such choices is $F(N)^{j^2}$. Clearly any admissible jN by jN square is included, since it must satisfy the constraints within the N by N squares. But some outcomes are not admissible because they violate constraints across the small squares. Thus

$$\log F(jN)/(jN)^2 \leq j^2 \log F(N)/(jN)^2 = \log F(N)/N^2, \quad j \in \mathcal{N}_+,$$

giving

$$C \leq \log F(N)/N^2. \tag{7.2}$$

Now place the j^2 N by N squares on a square of size $jN + (j - 1)b$ such that each pair of squares is separated by b unspecified symbols. If, for any N and fixed b, and arbitrary choices of the admissible configurations of the N by N squares, it is possible to satisfy the constraint with some choice of the unspecified symbols, the combinatorial entropy is lower bounded by

$$C \geq \log F(N)/(N + b)^2. \tag{7.3}$$

This is the case since all the outcomes satisfy the constraint. If there is a finite b such that the condition of the lower bound is satisfied, the two bounds converge for increasing n, and consequently they converge to the combinatorial entropy. However, there are some rather simple cases in which it is not possible to apply the lower bound.

There is no simple way of counting admissible configurations in two dimensions, and, even for small N, the enumeration can be a tedious task.

Example 7.4 (Counting hard-square configurations).　*For the hard-square constraint there are 63 configurations on a 3 by 3 square. Thus we can get an upper bound on the combinatorial entropy as $C < \log(63)/9 = 0.66$. The squares can be combined to give admissible configurations by including a border of zeros, $b = 1$. Thus the lower bound becomes $C > \log(63)/16 = 0.38$.*

Methods to achieve tighter bounds (for given limited computational resources) are introduced below.

7.4   Bands of finite height

The analysis for 1-D sequences introduced in Chapter 2 is readily generalized to *bands*, defined as 2-D arrays of finite fixed (vertical) height M and arbitrary (horizontal) width n. These bands are sequences of k-dimensional vectors, thus extending the alphabet from $\mathcal{A}$ to $\mathcal{A}^k$ representing k elements.

Multi-track storage systems using a number of parallel tracks (often with a hexagonal arrangement of the symbols) can provide increased capacity on DVD-like disks. Thus bands are used as an intermediate technology that achieves greater density than that of 1-D tracks. Here we use these bands as an approximation to 2-D fields and the analysis of finite bands lets us derive bounds on the entropy of 2-D fields.

The admissible configurations of an array of height M may for all n be described by a finite-state source and the transfer-matrix approach may be used to count the number of configurations.

For a constraint of finite extent (L), the states of the source are given by the symbols on the M by $L - 1$ segment which appear as the first or last $L - 1$ columns of an admissible configuration on an M by L rectangle, i.e. a configuration of $E(M, L)$. A transition from state i to state j is admissible if there is a configuration in $E(M, L)$ for which state i is identical to the left $L - 1$ columns and state j to the right $L - 1$ columns. States i and j have an overlap of $L - 2$ columns. The last column of j is generated by the transition from i to j and appended to the previous columns of the output. Any admissible configuration of $E(M, n)$ with fixed M and n (with $n > L - 1$) columns may be generated as an output by starting the source in the state specified by the first $L - 1$ columns and making $n - L + 1$ transitions appending one column to the output in each transition. The transfer matrix $\mathbf{T}_M$ indicates which transitions of the band of height M satisfy the constraint by defining the elements $t_{ij} = 1$ if the transition from state i to j is admissible and $t_{ij} = 0$ if it is not admissible.

This transfer matrix may be used to express the number of configurations on rectangles with the height of the band. For a given transfer matrix $\mathbf{T}_M$, the number of configurations on the band of height M after n transitions is, as in the 1-D case, expressed by $\mathbf{u}'\mathbf{T}_M^n\mathbf{u}$, where $'$ denotes the transpose.

The combinatorial entropy per symbol of a band of height M is

$$\frac{h(M)}{M} = \frac{\log \Lambda_M}{M}, \tag{7.4}$$

where again Λ_M is the largest (positive) eigenvalue of $\mathbf{T}_M$. By the same subadditivity argument as used for counting configurations on a rectangle, $h(M)/M$ is also an upper bound on the combinatorial entropy of the field.

Again as for the rectangles, for constraints such that any two configurations, X and Y, on arrays of height M may admissibly be concatenated (or cascaded) by padding a merging array, V, of c rows to form the admissible configuration XVY, the capacity is lower bounded by $h(M)/(M + c)$. We refer to such constraints as being band *mergeable*.

Figure 7.2. Domino tiling of the plane ([4], ©1999 IEEE).

Thus we have the bounds on the combinatorial entropy

$$\frac{h(M)}{M} \geq C \geq \frac{h(M)}{M+c}, \tag{7.5}$$

where the lower bound applies only to mergeable constraints with merging height c. The minimum-distance constraints are obviously mergeable.

Example 7.5 (Merging hard-square bands). *The 2-D RLL $(1, \infty)$ is mergeable. Any two admissible arrays may be merged by padding one row of all 0s, i.e. $c = 1$. For $M = 20$ and using (7.5) we get the result $0.5928 \geq C \geq 0.5646$.*

Example 7.6 (Tiling of the plane with dominoes). *Dominoes are tiles of 2 by 1 elements (possibly rotated by 90 degrees; see Fig. 7.2). Requiring the plane to be tiled with dominoes is not a mergeable constraint. Consider a, say, horizontal edge of vertical tiles having alternating offset. Extending the field from the edge is predetermined except at the end. For any chosen width between two arrays the two opposing edges of which do not lock up over a longer stretch than this width, there is no merging array that can merge these two arrays in accordance with the constraint.*

7.4.1   Bounds for first-order-symmetric constraints

For constraints with symmetric transfer matrices, e.g. the hard-square constraint, it is possible to derive some good upper and lower bounds on the combinatorial entropy. The approach taken is to bound the largest eigenvalue of the transfer matrix (7.4). For a symmetric constraint defined on an $L^2 = 2 \times 2$ square the transfer matrix is symmetric and we refer to this as a first-order-symmetric constraint.

First we introduce a modification to the band. Consider a band of height m (and arbitrary width n), now place it on a *cylinder* such that, in each column, elements with indices 1 and m become neighbors. The transfer matrix will be modified, but the entropy of the cylinder may be expressed on the basis of the new transfer matrix as for the bands.

The approach taken will be to consider the number of configurations on rectangles. Initially n is fixed and a bound on the number of configurations for $m \to \infty$ is expressed. This bound is given by an expression for a finite m. For fixed m, we can then define a finite-state source in the n-direction and let $n \to \infty$.

For the upper bound, consider the transfer matrix $\mathbf{T}_n$ for a band extending in the vertical direction. We shall apply the result that for any real symmetric matrix, $\mathbf{T}$, the largest eigenvalue Λ satisfies

$$\Lambda^{2p} \leq \text{Trace}(\mathbf{T}^{2p}),$$

since the eigenvalues are real and thus positive when squared and the trace equals the sum of the eigenvalues. Applying this to $\mathbf{T}_n$ and solving for the largest eigenvalue Λ_n gives

$$\Lambda_n \leq \text{Trace}(\mathbf{T}_n^{2p})^{1/2p} \tag{7.6}$$

for all positive integers p.

Thus the eigenvalue which determines the growth rate of the number of configurations for fixed n and large m is upper bounded by the trace of $\mathbf{T}_n^{2p}$. It may be observed that the trace counts the number of configurations starting and ending in the same state after $2p$ transitions. We may interpret these as solutions on a cylinder of finite circumference $2p$ and extending n elements horizontally, satisfying the constraint all the way around the cylinder as introduced above. Keeping $2p$ fixed, we can now define a finite-state source with states of width $L - 1$ on the cylinder and use the cylinder entropy to express the limit for $n \to \infty$:

$$C = \lim_{n \to \infty} \log \Lambda_n \leq \lim_{n \to \infty} \log \text{Trace}(\mathbf{T}_n^{(2p)})^{1/(2p)} \leq \frac{h'(2p)}{2p}, \tag{7.7}$$

where $h'(2p)$ is the capacity of the cylinder with circumference $2p$.

A lower bound on the capacity for the RLL $(1, \infty)$ problem is derived using bands in the m-direction followed by a limit in the n-direction. Again we consider the transfer matrix for a band of width n, $\mathbf{T}_n$.

We need a general result from linear algebra. Since the transfer matrix is symmetric, the largest eigenvalue, Λ, is bounded by

$$\Lambda_n \geq \frac{(\mathbf{y}, \mathbf{T}_n \mathbf{x})}{(\mathbf{y}, \mathbf{x})}, \tag{7.8}$$

where $(\mathbf{x}, \mathbf{y})$ denotes the inner product, i.e. $(\mathbf{x}, \mathbf{y}) = \mathbf{x}'\mathbf{y}$. (Sketch of proof: clearly the relation is satisfied with equality when $\mathbf{y}$ is a left eigenvector or $\mathbf{x}$ a right eigenvector of $\mathbf{T}_n$ corresponding to the largest eigenvalue; in other cases the vector can be expanded as a sum of eigenvectors, and the inequality follows.)

Further, we note that the number of configurations after $m - 1$ transitions is given by the inner product $(\mathbf{u}, \mathbf{T}_n^{m-1}\mathbf{u}) = \mathbf{u}'\mathbf{T}_n^{m-1}\mathbf{u} = F(m, n)$, which gives the number of configurations on an m by n rectangle. (Recall that $\mathbf{u}$ is the all-1s vector). This result could also be obtained by introducing an adjacency matrix, $\mathbf{T}_m$, in the other direction, i.e. $(\mathbf{u}, \mathbf{T}_m^{n-1}\mathbf{u}) = F(m, n)$. Since the constraint is symmetric, $\mathbf{T}_n = \mathbf{T}_m$ for $n = m$.

On replacing $\mathbf{T}_n$ by $\mathbf{T}_n^p$ and letting $\mathbf{x} = \mathbf{y} = \mathbf{T}_n^q\mathbf{u}$, we get

$$\Lambda_n^p \geq \frac{(\mathbf{T}_n^q\mathbf{u}, \mathbf{T}_n^p\mathbf{T}_n^q\mathbf{u})}{(\mathbf{T}_n^q\mathbf{u}, \mathbf{T}_n^q\mathbf{u})} = \frac{(\mathbf{u}, \mathbf{T}_n^{p+2q}\mathbf{u})}{(\mathbf{u}, \mathbf{T}_n^{2q}\mathbf{u})} = \frac{F(p+2q+1, n)}{F(2q+1, n)}, \tag{7.9}$$

Since $\mathbf{T}_n$ is symmetric and thereby identical to its transpose.

The numerator and the denominator may be interpreted as the numbers of configurations after $p + 2q$ and $2q$ transitions, respectively. This is expressed by $F(p + 2q + 1, n)$ and $F(2q + 1, n)$ on the corresponding rectangles. We can now keep p and q fixed, and let $n \to \infty$. Thus we have bands of width $p + 2q + 1$ and $2q + 1$ with adjacency matrices $\mathbf{T}_{p+2q+1}$ and $\mathbf{T}_{2q+1}$, respectively. The limits are given by the band entropy, i.e. $\lim_{n\to\infty} F(m, n) = h(m)$. Taking the logarithm, we may express the bound on the combinatorial entropy by band entropies,

$$C \geq \frac{h(p + 2q + 1) - h(2q + 1)}{p}. \tag{7.10}$$

Example 7.7 (Bounds for the hard-square constraint). *The bounds, (7.7) and (7.10), above provide very tight bounds for the RLL $(1, \infty)$ constraint: $0.587\,891\,161\,7 \geq C \geq 0.587\,891\,164$. These bounds were obtained with $M = 20$ $(h(20) - h(19))$ for the lower bound and $M = 18$ $(h'(18))$ for the upper bound.*

The methods apply to other (first-order-symmetric) constraints, but they do not in general apply when $L > 2$. Experience suggests that taking the difference between the entropies of two bands of different heights as in (7.10) does indeed provide a very good estimate of the entropy. This may loosely be motivated by saying that the boundary effects are reduced and the difference expresses the entropy within the interior. There has to date been no analysis of this estimate.

7.4.2   Markov chains on bands

As in the simple 1-D case, we can assign probabilities to the state transitions in a band, and in this way we obtain a description in the form of a Markov chain. The transition probabilities p_{ij} must be chosen such that $p_{ij} \geq 0$ if $t_{ij} > 0$, $p_{ij} = 0$ if $t_{ij} = 0$, and $\sum_j p_{ij} = 1$. Let $\mathbf{P}_m$ denote the transition-probability matrix corresponding to a given $\mathbf{T}_m$.

We can then proceed to develop an encoding of information as discussed for simple sequences, and, by varying the probabilities, the entropy can be maximized to get the value of the combinatorial entropy. If a truly 2-D source is processed m rows at a time, we can use 1-D encoding techniques. The band entropy can be maximized with a given state distribution, and in this way the difference between the rows inside the band and along the edges can be reduced.

7.5 Causal models in two dimensions

An essential difference between data sets in one and two dimensions is that there is no unique natural sequencing of the symbols in two dimensions. Two-dimensional models were introduced in Chapters 2 and 5, and here the concepts are further developed and treated. In some applications, the possibility of exploiting the structure in different directions is one of the advantages of representing data in this way. However, for purposes of coding and other forms of signal processing, it may be necessary to select a particular order.

7.5.1 Basic concepts

Assume that the symbols on the M by N rectangle have been assigned indices $1, \ldots, MN$. It would be convenient to have the symbols in the neighborhood of symbol j, $W(j)$, have indices close to j. However, no such mapping of 2-D space onto one dimension is possible. In the common processing order, rows are scanned one at a time, from left to right, and the original indices (i, j) are mapped onto $(N - 1)i + j$.

A processing method that uses a single index is called causal, since it allows us to distinguish "past" and "future" symbols. In particular, the neighborhood $W(j)$ is separated into a past (known) part $U(j)$ with index j, and a future (unknown) part.

We can now write the chain rule for the entropy, $H(X)$, of the M by N rectangle using the single index as

$$H(X) = \sum_{j=1}^{MN} H(x_j | x_1, x_2, \ldots, x_{j-1})$$

and, since the entropy does not decrease when some of the conditions are suppressed, we may write

$$H(X) \leq \sum_{j=1}^{MN} H(x_j | U(j)). \tag{7.11}$$

Thus, if we can estimate the probability distribution of the context U in a given data set, and similarly find the conditional distribution of the next symbol, we can find an upper bound for the entropy. However, even if the field is a Markov random field (MRF), and the distribution of x_j thus depends only on $W(j)$, the bound usually does not give the true entropy. The reason for this is that the unknown part of $W(j)$ depends to some extent on the known outcome outside $U(j)$. In one dimension we get the true entropy of a Markov chain by applying the chain rule, but in that case the entire future outcome depends on the past only through the current state. In two dimensions the future depends on the entire boundary of the known set, which for finite constraints is the last $M - 1$ rows (in a row-by-row traversal). We can get a better approximation by using a larger template, including a longer segment of the last $M - 1$ rows both to the left and to the right. However, it is difficult in general to analyze the convergence of such an approach.

7.5.2 Bit-stuffing in two dimensions

Bit-stuffing is a simple coding technique that may also be applied e.g. to 2-D RLL (d, ∞) constraints, where a 0 may always be written. Consider the problem of mapping an input string of information symbols onto a 2-D configuration admissible by the given constraint. The 2-D field is traversed in a predefined order, say row by row. As in the 1-D case, each time a new information symbol is written, one or more new symbols may be uniquely determined by the constraint and the causal elements of the field. If this is the case, the so-determined symbols are stuffed, i.e. written at the given positions.

In the basic version the input string is simply the information sequence to be coded, which we often assume to be an i.i.d. binary sequence. In a more elaborate coding scheme, a precoding step may be applied to map the information sequence onto an input string having biased probabilities. For the RLL $(1, \infty)$ constraint, a bias toward writing a 0 at the free positions seems advantageous, since each new 1 will force two future elements to be 0. The bit-stuffing scheme may be described by a sequence of conditional probabilities. When a symbol is uniquely determined by the constraint and the causal elements, the so-determined symbol which is stuffed is assigned the probability 1 (and the so-forbidden symbol(s) is (are) assigned probability 0). When we are free to choose, the (conditional) probability is given by the probabilities of the biased input string in bit-stuffing. Thus the so-called bit-stuffing is a context-based model as described in Chapter 5.

In the case of bit-stuffing for the 2-D RLL $(1, \infty)$ constraint, the probabilities will be conditional on the previous symbol and the symbol above. If one of these elements has the value 1, a 0 is stuffed. This approach will be analyzed in the next section, and the analysis will give an example of increasing the entropy by using a biased input sequence.

For constraints for which it is always possible to write, say, a 0, bit-stuffing provides a means of coding. The minimum-distance (between 1s) constraints constitute such a class of 2-D constraints.

7.6 Pickard random fields

In this section we study a special type of random field that lets us explicitly calculate the probability distribution of finite sets of symbols, and thus we can also calculate the entropy. The motivation for studying these fields in particular is that we avoid the difficulties discussed in the previous section. An important property is that the rows and columns have distributions described by Markov chains. The Pickard random fields (PRF) were introduced in Chapter 2 in a simple version. Here the general version shall be introduced.

The short notation for the variables is used. Thus A, B, C, and D represent the four variables within a 2 by 2 square located at (i, j):

$$\begin{bmatrix} X_{ij} & X_{i,j+1} \\ X_{i+1,j} & X_{i+1,j+1} \end{bmatrix} = \begin{bmatrix} A & B \\ C & D \end{bmatrix}.$$

Example 7.8 (Bit-stuffing for the hard-square constraint). *In Chapter 2, PRFs were considered for the hard-square constraint. If the pixel above (B) or the pixel to the left (C) is 1, the current pixel (D) must be 0, i.e. stuffed. The maximum entropy for the hard-square PRF was obtained by selecting $P(D|ABC) = P(D|BC)$ given by $P(D = 1|B = C = 0) = 4/13$, which yielded $H(D|ABC) = 0.5788$. Thus a bias toward selecting the 0 symbol increases the entropy.*

7.6.1   General Pickard random fields

In the PRF introduced in Chapter 2, assumptions were made about the conditional independence of diagonal elements, and a single Markov chain described the distribution of any row (from left to right) and any column (from the bottom up). Here the concept is generalized such that the rows and columns may be described by two different Markov chains, one chain for the rows and one chain for the columns. Thus only data with such a simple structure can be described, and, in terms of finite constraints, only the case $L = 2$ is covered.

A requirement of a PRF is that there exist *stationary* probabilities on 2 by 2 elements denoted A, B, C, and D. This may be expressed by the following properties of the distribution $(ABCD)$:

$$P(A) = P(B) = P(C) = P(D),$$
$$P(AB) = P(CD),$$
$$P(AC) = P(BD).$$

The third condition implies that the conditional probability $P(BD|AC)$ defines a Markov chain on two rows. Likewise $P(AB)$ and $P(A) = P(B)$ define a Markov chain, which, by virtue of the identity, will be the same for CD. The challenge is to make these Markov chains consistent such that the two one-row Markov chains are marginals of the two-row Markov chain.

Now, the first step in the construction is to represent each row as an irreducible Markov chain. Thus, for an alphabet of $|\mathcal{A}|$ symbols, the chain has $|\mathcal{A}|$ states with stationary probabilities P_j and transition matrix $\{q_{ij}\}$.

Any irreducible Markov chain can be reversed by a simple operation as discussed in Section 2.3.2, and, in many of the cases we are interested in, the chain is symmetric with respect to such a reversal.

To arrive at a PRF, we need two extra conditions, which make each of the rows Markov chains with the same given transition matrix. Thus the stationary probability distribution on a pair of variables in a column, x_{ij} and $x_{(i+1)j}$, is given, and we want a transition matrix that specifies the next pair, $x_{i(j+1)}$ and $x_{(i+1)(j+1)}$. We again use A, B, C, and D to denote these variables.

The first condition is the independence assumption

$$P[B|AC] = P[B|A]. \tag{7.12}$$

This condition makes row i a Markov chain, since the conditional probability of each variable now depends only on the symbol to the left, and we can specify the distribution on the rest of row i by repeating this rule. Similarly, we can make row $i + 1$ a Markov chain by requiring that

$$P[C|BD] = P[C|D]. \tag{7.13}$$

The argument is the same, but now considering row $i + 1$ of the reverse two-row Markov chain. Note that, to make the two rows equal, we use the reverse Markov chain in this case, which is consistent with the stationarity.

Thus repeating the construction row by row yields a stationary field, where rows are described by the same Markov chain. That this also applies to columns may be seen by applying the independence assumptions (7.12) and (7.13) to the two-column Markov chain.

A stationary distribution $(ABCD)$ satisfying (7.12) and (7.13) defines a PRF. The boundaries are given by the Markov chains $P[B|A]$ horizontally and $P[C|A]$ vertically. The conditional probability $P[D|ABC]$ is given by $(ABCD)$.

The independence assumption (7.13) may also be replaced by the combination of

$$P[A|BD] = P[A|B] \qquad \text{and} \qquad P[A|CD] = P[A|C],$$

which is the same as (7.12) and (7.13) in the reverse directions.

In Chapter 2 the two Markov chains of the PRF were chosen to be identical. Together with the independence conditions (7.12) and (7.13), $P[CAB]$ and $P[BDC]$ are found as the probabilities of three consecutive symbols in the same chain. This is clearly consistent with the stationarity condition that AC and BD have the same distribution. If we want to construct a PRF for a 2 by 2 constraint, then, starting with a Markov chain we must find a distribution for $ABCD$ which satisfies the 2 by 2 constraints and also is consistent with the marginal distributions on ABC and BCD. Such a distribution must satisfy a system of linear equations. For any pair of values, $(C = c_i, B = b_j)$, the probability can be found by summing the ABC distribution over A (or the BCD distribution over D, with the same result). The distribution $(A = a_r, C = c_i, B = b_j, D = d_s)$, which includes at most $|\mathcal{A}|^2$ nonzero terms, can be summed over A and D to give at most $2|\mathcal{A}|$ equations.

Not all Markov chains will produce positive solutions to these equations, and, for some 2 by 2 constraints, the distribution of three variables cannot even be expressed as a Markov chain. However, in other cases we arrive at a convenient description of a field satisfying the constraint, and there are often parameters that can be chosen to match desired properties or to maximize the entropy.

In Section 7.8.1 we shall return to the problem of constructing a PRF for a constrained field.

7.6.2 Remarks on causal models

In any MRF defined by the interaction on 2 by 2 elements $(L = 2)$, the probabilities of the variables below row j depend on the upper part of the plane only through row j. Thus the field can be constructed by specifying the probability distribution of two rows, j and

$j + 1$, such that the two rows have the same distribution. We can then find the probability of row $j + 1$ given row j, and this conditional distribution can be used to specify each subsequent row given the one above. It is easy to describe two rows by a Markov chain with at most $\mathcal{A}^2$ states, which was an essential part of the PRF construction. In general, the two rows will have the same distribution if the two-row Markov chain is symmetric in the two variables, and the transitions can, in particular, be chosen to satisfy a symmetric constraint with $L = 2$. However, such an approach will usually not make the two rows Markov chains (unless restrictions such as the independence assumptions (7.12) and (7.13) are imposed). They are hidden-state Markov sources with the variable in the other row as a hidden-state variable. The entropy is not easily calculated, though it may be bounded from both sides by (2.9) and (2.10). For more than two rows, it is much more difficult to define a Markov chain on multiple rows yielding a stationary distribution.

As a consequence of the description in terms of conditional entropies, the entropy of the field can be maximized by expressing each row as a distribution conditioned on the previous $L - 1$ rows. If all rows have the same probability distribution, the problem can then be shifted down one row, and the field is generated recursively one row at a time, such that any L rows are drawn from the same distribution. The entropy may be expressed by

$$H(X) = H_L - H_{L-1},\qquad(7.14)$$

where H_L and H_{L-1} are the entropies per column for L and $L - 1$ rows, respectively. There are two difficulties in actually applying this approach in the construction of the field: in general it is difficult to assure that the rows have the same distribution, and even for simple constraints there is no simple description of the row distribution that will maximize the entropy.

An indirect way of finding a row distribution that approximates the maximum-entropy distribution and also assures the same distribution for the rows is based on the cylinder descriptions introduced previously. Imposing rotational symmetry on the cylinder directly leads to an identical distribution of the rows. For a cylinder of circumference $D > L$, let the distribution on L rows be defined by this cylinder with $D - L$ rows being hidden. Defining the next row conditional on the $L - 1$ previous rows recursively one row at a time gives a stationary distribution. For a cylinder of sufficiently large circumference, the distribution is close to that of the maximum-entropy field. The entropy can be expressed as the difference (7.14), where the entropies are given by the entropies of a function of a Markov chain. These entropies are not easy to calculate, but they may be bounded from above and below by (2.9) and (2.10).

7.7 Markov random fields with maximum entropy

The 2-D generalization of a Markov chain is called a Markov random field (MRF). We do not give a general discussion of this concept, since it leads to analytical problems that usually cannot be solved. However, the following discussion gives the general flavor of the concept and demonstrates in particular that a stochastic description with entropy equal to the combinatorial entropy could be obtained in this way. Thus, quite apart from analytical and practical problems, the MRF is a very interesting concept for 2-D constrained fields and optimization of the entropy.

An MRF is defined by a conditional probability assignment on any finite set of symbols conditioned on a sufficiently wide boundary enclosing these symbols. The Markov condition in two dimensions says that the probability distribution on a finite set of symbols conditioned on the rest of the plane equals the probability distribution conditioned on a finite boundary consisting of all symbols that are neighbors of the given set. For fields defined by finite constraints, it is sufficient to specify a boundary that is $L - 1$ symbols wide, thus separating the past and the future.

The MRF with maximum entropy is defined by assigning equal probability to all admissible outcomes conditioned on the boundary. In order to verify that this is in fact a consistent definition of probabilities, we note that, if certain symbols are specified, leaving a subset of the original set, the outcomes on the subset still have equal probabilities. It is clear that, as we make the area larger, we can make the conditional entropy approach the combinatorial entropy. However, there are some serious difficulties associated with this definition. First of all, we are again left with the problem of counting outcomes. This number is a special case of a function known as the partition function for MRFs, and the difficulty of finding this function is the main drawback of the theory. The other problem is that knowing the conditional probabilities does not give us a tool for calculating the probability of any specific finite configuration of symbols (or even a single symbol). Thus, from a coding perspective, MRFs are useful mostly as a theoretical concept.

7.8   Block Pickard construction

For larger constraints ($L > 2$) and causal models with contexts extending beyond the three nearby PRF context variables, we generalize the construction of the previous section by using blocks of symbols. Thus several symbols are collected into a block, which is treated as a super symbol, and the plane is covered with such blocks arranged in a suitable grid.

Example 7.9 (No isolated bits (n.i.b.)).   *The following constraint on binary variables has been suggested as a method to reduce the effect of interference between neighboring symbols. All configurations of binary symbols are admissible except*

$$
\begin{array}{ccc}
& 0 & \\
0 & 1 & 0 \\
& 0 &
\end{array}
\qquad \text{and} \qquad
\begin{array}{ccc}
& 1 & \\
1 & 0 & 1 \\
& 1 &
\end{array}
$$

since in these two cases the combined effects of the neighbors might cause a wrong detection of the central symbol. Thus we have a constraint with $L = 3$, but diagonal neighbors are not important; the block Pickard field can be constructed with blocks of 1 by 2 elements (see Fig. 7.3). We consider four neighboring blocks, A, B, C, and D, and

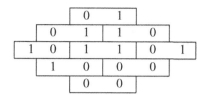

Figure 7.3. An example of an n.i.b. configuration on a 3 by 3 field.

index the elements of these by subscripts,

$$c_1 c_2$$
$$a_1 a_2 d_1 d_2$$
$$b_1 b_2$$

Given A, B, and C, we must choose D in such a way that neither a_2 nor d_1 becomes isolated. Also we notice that, if d_1 were chosen to make a_2 isolated or d_2 were chosen to make d_1 isolated, this would involve only (causal) elements in A, B, and C. So, when selecting D, it is sufficient to check A, B, and C, i.e. a PRF assigning probability 0 to any forbidden configuration of $ABCD$ will also do so for all forbidden configurations extended to multiple blocks by the PRF.

7.8.1 Constructing Pickard random fields from two Markov chains

In Chapter 2, a method for deriving a PRF from one Markov chain was presented. This is generalized to the case in which the two Markov chains along rows and columns are different. These Markov chains also define the top and left boundaries.

It is essential for the construction that the stationary distribution on the states is the same for the two processes. However, as discussed above, the pairs AB and AC in general have different distributions. Nevertheless, we recall the Pickard independence assumptions (7.12) and (7.13)

$$P[B|AC] = P[B|A], \qquad P[C|BD] = P[C|D]$$

and note that the transitions between neighboring elements are transitions of a Markov chain and that the triples CAB and CDB are given by transitions of the two Markov chains but in reverse order.

Thus the distribution on BC can then be found in two ways, starting with either a horizontal or a vertical transition. If the two transition matrices for the block processes are **R** and **S**, and the distribution of B is given by a vector $\mathbf{p}(B)$, we can find the distribution of C by multiplying by either **RS** or **SR**. The only way we can assure consistency is by requiring that the transition matrices commute,

$$\mathbf{RS} = \mathbf{SR}. \tag{7.15}$$

For $\mathbf{R} = \mathbf{S}$, as in Chapter 2, the requirement is obviously satisfied. When **R** and **S** differ, the constraint can be written out as a set of nonlinear equations in the entries of **R** and **S**. If two commuting n by n matrices have n distinct eigenvalues and n eigenvectors, the eigenvectors are the same. In particular, this means that the two processes have the same stationary distributions.

Once a pair of transition matrices has been obtained, we complete the construction as outlined previously by solving a system of linear equations for each pair of values for BC to get the $ABCD$ distribution. There are clearly more variables than equations, but the equations might not have a solution. At this stage we limit the terms of the $ABCD$ distribution to those configurations that satisfy the constraints. We have tested earlier that all combinations of ABC and BCD are admissible, but the squares that include variables from both A and D have yet to be tested.

Since the number of variables tends to be large, we prefer an approach that also maximizes the entropy. Let the values of B and C be fixed and compute the marginal distribution on CAB and CDB from the Markov chains. We can then write the probability distribution of $ABCD$ as a matrix with the value of A as the row index and the value of D as the column index. The sum of the rows must equal the distribution on BCD and the sum of the columns is the ABC distribution. We initialize the matrix with zeros in positions that are not admissible by virtue of the constraint and 1s in the remaining places. In an iterative process known as *iterative scaling*, we modify the matrix $\{m_{ij}\}$ to get row sums r_i and column sums c_j. In each step we update the rows

$$m'_{ij} = m_{ij} \times r_i \Big/ \sum_j (m_{ij}) \tag{7.16}$$

and subsequently the columns are updated,

$$m'_{ij} = m_{ij} \times c_j \Big/ \sum_i (m_{ij}). \tag{7.17}$$

If the linear equations have a non-negative solution, the iteration will converge to it. We omit the analysis of the general case, but the idea is suggested by considering the binary case in the following example.

Example 7.10 (Iterative scaling).   *Find the probability distribution on A and B when they cannot both be 1, $P[A = 1] = a$, $P[B = 1] = b$, $P[AB = 00] = c$. In the 3-D (x, y, z) space the A distribution $x = a$, $y + z = 1 - a$ is a line, as is the B distribution $y = b$, $x + z = 1 - b$. The obvious solution is the intersection of the lines, $z = 1 - a - b$. It is clear from geometric considerations that we can move toward the solution starting from a point with positive coordinates by scaling each coordinate by a positive factor in such a way that the point is shifted to one of the lines. As we repeat the scaling, the point is moved alternately to one and the other line until it reaches the intersection. If the lines did not intersect, we would end up alternating between two points, and this would also be the case if the intersection had a negative coordinate $(a + b > 1)$.*

In general the two marginals define two linear subspaces of the distribution space. We assume that they intersect in a subspace that includes points with positive coordinates. Usually the solution is not unique, but depends on the initial point.

The result of the iterative scaling is a matrix in which each entry satisfies

$$m_{ij} = t_i s_j,$$

where t_i is a constant that depends only on the row index and s_j depends only on the column index. In Appendix B we present an analytical approach to the maximization and demonstrate that it leads to a solution of this type.

The application of iterative scaling to determine the distribution on $ABCD$ of a PRF for a 2-D constraint given two Markov chains for which the matrices commute is given explicitly below. For each configuration bc, the elements are given by

$$m_{ij} = P(A = i, D = j|bc), \quad (i, j) \in \mathcal{A}^2, \tag{7.18}$$

where $m_{ij} = 0$ if the corresponding configuration $abcd$ is forbidden. The row and column sums are given by

$$r_i = \sum_{d \in \mathcal{A}} P(A = i, D = d|bc), \quad i \in \mathcal{A}, \tag{7.19}$$

$$c_j = \sum_{a \in \mathcal{A}} P(A = a, D = j|bc), \quad j \in \mathcal{A}. \tag{7.20}$$

Iterative scaling is applied for each bc pair. The initial distribution, $Q(AD|BC)$, is set to a uniform distribution over the admissible configurations $abcd$ for each bc (and to 0 for the forbidden configurations). For each bc a sequence of distributions is generated by iterating (7.16) and (7.17), converging to the solution $P^*(AD|BC)$ in the limit achieving the row and column sums $P(bcd)$ and $P(abc)$, respectively.

The entropy of the interior, $H(D|ABC)$, may be expressed by

$$H(D|ABC) = H(ABCD) - H(ABC)$$

$$= H(BC) + H(AD|BC) - H(ABC), \tag{7.21}$$

where $H(BC)$ and $H(ABC)$ are given by the Markov chains defining the boundary and $H(AD|BC)$ is given by the distribution determined by the iterative scaling. Thus optimizing $H(AD|BC)$ will optimize $H(D|ABC)$ in the given setting.

Iterative scaling can also be used to find a solution to the other version of Pickard fields mentioned in Chapter 2. Starting from an initial distribution on $ABCD$ that is positive for all permissible configurations, find the marginal distribution for any of the known values, $ABC = abc$, etc., and scale the contributions to give the required values. By alternating between the known values for ABC, ACD, and BD (which must have the same distribution as AC), the distribution can be made to converge if a solution exists.

If we take an initial distribution, $Q(x)$, and apply iterative scaling, assuming that there is a non-empty class of distributions that will solve the equations, we find a distribution, P^*, which is closest to Q (in the sense that it minimizes the *divergence*, $D(P^*\|Q)$, where $D(P \| Q) \equiv \sum P(x) \log(P(x)/Q(x)))$. Starting with a uniform initial distribution over the admissible configurations such as Q maximizes the entropy, $H(P)$, over the distributions P which are solutions to the linear equations.

7.8.2   Block Pickard random fields for two-dimensional constraints

As in the example for the n.i.b. constraint, we introduce blocks that solve the equations, organized along the diagonals as a step toward designing block Pickard random fields for 2-D constraints. For an L by L constraint in general, if any L by L square containing an element in D contains only elements in A, B, and C, a block PRF can check the constraint on the elements it generates. Further, we are interested in avoiding L by L squares with elements in both B and C, since this could conflict with the independence of B and C given A (7.12). To address these issues, let the blocks be rectangles of size $L - 1$ by $2(L - 1)$ symbols, and let them be arranged in rows, but in such a way that each row is shifted by half a block relative to the previous row (see below). The blocks are identified by two indices, the first one indicating the position on a diagonal from upper left to lower right. Thus block $g(i, j)$ includes rows $(i - 1)(L - 1) + 1$ to $i(L - 1)$. The other index increases along the diagonal from lower left to upper right.

We propose a field, which is described by a causal model in which the blocks are specified one diagonal at a time starting from the upper left and extending to the lower right. Each diagonal is given conditioned on the one below it and to the left. If we need to refer to a specific variable, it will be given a third index numbered $1 \ldots 2(L - 1)^2$ in the same way as within each block.

Consider blocks $g(i, j)$, $g(i + 1, j)$, $g(i, j + 1)$, and $g(i + 1, j + 1)$, which for simplicity are again labelled $ABCD$, respectively, and the individual variables are given by lower-case letters in the diagram below.

The block structure introduced above is illustrated in the diagram below for $L = 3$:

$$c_1 c_2 c_3 c_4$$
$$c_5 c_6 c_7 c_8$$
$$a_1 a_2 a_3 a_4 d_1 d_2 d_3 d_4$$
$$a_5 a_6 a_7 a_8 d_5 d_6 d_7 d_8$$
$$b_1 b_2 b_3 b_4$$
$$b_5 b_6 b_7 b_8$$

With this choice, the constraints on symbols in D involve only the three given previous blocks, blocks A and B in the previous diagonal, and block C, which is the previous block in the current diagonal.

The transition between A and B (or C and D) is described as a transition in the process along the main (upper right to lower left) diagonal, whereas the transition in the other process goes from A to C, i.e. from $g(i, j)$ to $g(i, j + 1)$. Thus the transitions CAB and CDB consist of one transition in each process.

The block processes along the diagonals can be expressed as Markov chains since the constraints on a particular block cover only parts of the previous block. Also no constraint involves both B and C, and thus it is not impossible off-hand that they are independent. The construction of the Markov processes is similar to the construction of band processes in an earlier section: each admissible block represents a state of the process, and the possible state transitions are found by applying the L by L constraint to all positions that include symbols in two consecutive blocks. However, note that in the present construction the states do not overlap. In this context we test positions where some variables within the L by L square are not specified. Such configurations are considered admissible if there is a choice of the unspecified variables that makes the configuration admissible. We could have adopted the same approach when applying constraints to finite rectangles or bands, but in these cases it makes little difference. For the diagonal processes considered here there are no positions where all L by L symbols are specified, and not applying the constraint in this way would leave many impossible configurations to be eliminated later.

One of the difficulties of the construction is that, although the two processes can be chosen in a symmetric fashion, they usually cannot be identical. Thus we may choose to let two states have the same probability if they are symmetric with respect to a vertical axis. Similarly, the CD transition can be chosen to have the same probability as the CA transition when two states of block C are a reflected pair and the state of D is obtained by reflecting state A. However, if we consider the same non-symmetric states in the two processes, they represent different configurations, and it is quite possible that one is admissible while the other is not.

7.8.3 Block Pickard random fields – no isolated bit

The application of the block PRF construction to 2-D constraints is demonstrated by the 2-D no-isolated-bit (n.i.b.) constraint (Fig. 7.3). The diagonal processes have four states, and all transitions are possible.

Example 7.11 (Variable-to-fixed-length coding). *A simple solution for having the transition-probability matrices commute is simply to assign equiprobable transition probabilities to the transitions among the four symbols in the diagonal processes. In this case also the configurations on ABC and BCD become equiprobable. Assigning to $P(D|ABC)$ probabilities that are powers of 1/2 leads to a variable-to-fixed-length code. For each bc configuration consider the matrix with elements $m_{ij} = P(a(i)d(j)|bc)$. Let the indices of the blocks be given by the binary number of the binary symbols within the block, e.g. $a = 2a_1 + a_2, a_1, a_2 \in \{0, 1\}$ (Fig. 7.3). The row and column sums over m_{ij} must add to 1/4. For $bc = 00$, a solution is given below:*

$$m_{ij} = \frac{1}{4} \begin{pmatrix} 1/2 & 1/4 & 0 & 1/4 \\ 0 & 0 & 1/2 & 1/2 \\ 1/4 & 1/2 & 0 & 1/4 \\ 1/4 & 1/4 & 1/2 & 0 \end{pmatrix}. \tag{7.22}$$

Four of the 0 elements are due to the constraint, while the last 0 element, m_{33} (in italics), is required in order to find a solution restricted to powers of 1/2. The reason is as follows.

The row with two 0s precludes any column with all elements equal to 1/4. Having any column with (1/2, 1/4, 1/8, 1/8) requires two rows and two columns without 0 elements, which is not possible by virtue of the constraint for bc = 00. Thus we cannot have a column with four nonzero elements. By symmetry the solution above also provides the solution for bc = 33. The entropy per block for bc = 00 is $H(D|A, b = 0, c = 0) = 11/8$. In all other cases a solution may be found without introducing 0 elements other than those dictated by the constraint. These solutions are unique in terms of the entropy. For $P(D|ABC)$, the values 0, 1/2, 1/4, and 1/8 are used, thus coding 0 to 3 bits by each 2-bit block. For the boundaries the rate is 2 bits per symbol. For the interior the rates are $H(D|ABC) = 107/128 = 0.836$ bits per binary symbol.

In order to increase the entropy, we return to the more general setting of having different Markov chains along rows and columns. If we consider the transitions from C to A and D, it is clear that (10) to (00) are two different configurations in the two processes. However, if we consider the transitions from A to B and C, the configurations are obtained by reflection in the horizontal axis, and we could choose to assign the same probability to them. Here AB is the same transition as CD, but AC is a transition in the same chain as CA taken in the reverse order. Thus, in addition to symmetry around a vertical axis and symmetry with respect to an interchange of 0 and 1, we can simplify the construction by requiring that the two diagonal processes represent the same Markov chain with a reversal of the direction. (Unfortunately, that property does not imply that the transition matrices commute.) If we write the transition matrix (going from row to column) for the main diagonal as

$$\mathbf{R} = \begin{pmatrix} abcd \\ efgh \\ hgfe \\ dcba \end{pmatrix} \tag{7.23}$$

then the symmetry with respect to an interchange of 0 and 1 has been used to reduce the number of parameters. This property implies that the stationary distribution is $(1/2 - s, s, s, 1/2 - s)$. Since the rows of $\mathbf{R}$ sum to 1, we have

$$b + c = 1 - (a + d); \quad f + g = 1 - (e + h)$$

and, if s is used as a parameter, we get

$$(1/2 - s)(b + c) + s(f + g) = s,$$

which gives

$$f + g = 1 - (b + c)(1 - 1/(2s)).$$

If the other transition matrix, $\mathbf{S}$, is specified by using symmetry about a vertical axis, we simply have to interchange rows and columns 2 and 3 of $\mathbf{R}$. To make this matrix the transition matrix for the process in the main diagonal reversed, we multiply the rows of both by the stationary probability of the corresponding states, and then require that one of the results is the transpose of the other (this procedure amounts to verifying that $P[AB] = P[BA]$). This condition actually implies just that

$$e/c = (1/2 - s)/s,$$

since it follows from the previous relations that we then also have

$$h/b = e/c.$$

To make **R** and **S** commute, we calculate the product and notice that eight of the entries are already equal due to the symmetry of the matrices. The remaining eight products are equal if

$$(a - d - f + g)(b - c) = 0.$$

Example 7.12 (No isolated bits (n.i.b.) as a Pickard field with identical Markov chain boundaries). *Here we consider the simple case in which the diagonal process is symmetric, and thus the two Markov chains are identical. Further, let the four states have the same probability. Thus $b = c = e = h$ and $s = 1/4$. Let the transition matrix be*

$$\mathbf{R} = \begin{pmatrix} 2/9 & 2/9 & 2/9 & 1/3 \\ 2/9 & 2/9 & 1/3 & 2/9 \\ 2/9 & 1/3 & 2/9 & 2/9 \\ 1/3 & 2/9 & 2/9 & 2/9 \end{pmatrix}.$$

The entropy is increased by taking d, g larger than the other transition probabilities since these pairs are not part of any forbidden configurations. We can now find the value of $P[CB]$ for each of the 16 pairs of values, but because of the symmetry there are actually only five cases: $P[00, 00] = P[11, 11]$, $P[00, 10] = P[00, 01] = P[10, 00] = P[01, 00] = P[11, 10] = P[11, 01] = P[10, 11] = P[01, 11]$, $P[01, 01] = P[10, 10]$, $P[01, 10] = P[10, 01]$, and $P[00, 11] = P[11, 00]$. In the first three cases there are four, two, and three pairs of AD values that are not admissible; in the last two cases all pairs can occur. In all cases the sums of rows and columns give seven equations, since the row sums and the column sums add up to the same total. In the second case we find

$$P[C = 00, B = 01] = \sum_i P[C = 00, A = s_i, B = 01]$$

$$= 1/4(4/81 + 4/81 + 6/81 + 6/81)$$

$$= 5/81$$

and the four terms are also the row sums in the 4 by 4 matrix representing the $ABCD$ distribution. We get the same number by summing over D, and the terms (which in this case are the same) are the column sums of the distribution. The two cases that are not admissible are $(A, D) = (01, 00)$, $(01, 01)$, since the central 1 is isolated. Thus, with 1s *indicating admissible combinations, the distribution matrix is initialized as*

$$\begin{pmatrix} 1 & 0 & 1 & 1 \\ 1 & 0 & 1 & 1 \\ 1 & 1 & 1 & 1 \\ 1 & 1 & 1 & 1 \end{pmatrix}.$$

To find a solution by iterative scaling, we first scale the rows (taking out the common factor $1/324$):

$$\begin{pmatrix} 4/3 & 0 & 4/3 & 4/3 \\ 4/3 & 0 & 4/3 & 4/3 \\ 3/2 & 3/2 & 3/2 & 3/2 \\ 3/2 & 3/2 & 3/2 & 3/2 \end{pmatrix},$$

and subsequently scaling of the columns gives

$$\begin{pmatrix} 16/17 & 0 & 24/17 & 24/17 \\ 16/17 & 0 & 24/17 & 24/17 \\ 18/17 & 2 & 27/17 & 27/17 \\ 18/17 & 2 & 27/17 & 27/17 \end{pmatrix}.$$

By repeating the process the matrix is made to converge to

$$\begin{pmatrix} 1 & 0 & 3/2 & 3/2 \\ 1 & 0 & 3/2 & 3/2 \\ 1 & 2 & 3/2 & 3/2 \\ 1 & 2 & 3/2 & 3/2 \end{pmatrix}$$

and the conditional probabilities become $1/2$ when $A = 01$, and $1/4$ for the other values. In the first case the results are not rational numbers. When the initial matrix is the all-1s matrix, as in the last two cases, the result is the product of the marginal probabilities. After solving for the distribution in all five cases, we can calculate the entropy, $H(X)$, of the field as

$$H(X) = H[D|ABC] = \sum_{abc} P[abc]H[D|abc],$$

which becomes 0.8997. By varying the initial choices of (free) parameters in (7.23), we find the maximum for $s = 0.2241$, $a = 0.2250$, $d = 0.3360$, and $b = 0.2457$; $H(X) = 0.9157$.

7.8.4   Tiling with dominoes using Pickard random fields

Domino tiling of the plane was introduced in Example 7.6. Here we derive a Pickard field tiling the plane with dominoes. This is one of very few non-trivial 2-D fields for which the combinatorial entropy is known explicitly, and the numerical value is approximately 0.42. Nevertheless, certain configurations of dominoes give rise to long-range effects (Example 7.6), and it is not easy to get a random field with an entropy that approaches the theoretical value. Below a very simple field with a modest entropy is constructed.

The dominoes are represented as occupying two adjacent positions in a rectangular grid, and, depending on the orientation, they are labelled L, R, U, and W, for left, right, up, and down. Clearly an L is always followed by an R in a row, and a U by a W in a column.

The process is described by two Markov chains, one for rows (from left to right) and one for the diagonals from the lower left to the upper right. A state in each chain is a single symbol, and for the construction of the field we consider the configuration

$$\begin{pmatrix} - & A & B \\ C & D & - \end{pmatrix}.$$

Thus the diagonal is given by offsetting element C one column to the left relative to element A.

The offset is necessary in order to provide a means by which to look ahead in order to be able to avoid conflicts such as writing a left element forcing the next element to be a right element, which will induce a conflict if the element above this is defined as an up element.

Assuming that the states have the same stationary probability, we can write the row Markov-chain transition matrix with transitions from columns to rows as

$$\mathbf{R} = \begin{pmatrix} 0 & b+2c+d-1 & 1-b-c & 1-c-d \\ 1 & 0 & 0 & 0 \\ 0 & 1-c-d & c & d \\ 0 & 1-b-c & b & c \end{pmatrix},$$

where a, b, c, and d here denote transition probabilities. The transitions U to U and W to W are assigned the same parameter since they represent the same configuration of pieces. Similarly, for the Markov chain along the diagonal some of the transitions are the same, and are identified with row transitions. In particular, the diagonal transitions U to U and W to W are the same configuration as the row transition U to W (having probability b). In order to simplify the problem, we first assume that all diagonal elements in the transition matrix are the same, and we obtain

$$\mathbf{S} = \begin{pmatrix} b & e & c-e & 1-b-c \\ c & b & 1-b-c & 0 \\ 0 & 1-b-c & b & c \\ 1-b-c & c-e & e & b \end{pmatrix}.$$

Since the two matrices must commute, we compare the products of row 1 and column 4 and get the condition $d(c-e)=0$. As the least restrictive choice, we take $d=0$. It is now possible to make the matrices commute by taking $b=(1-c)^2$, $e=c(2b+2c-1)$.

As an easy numerical case we take

$$R = \begin{pmatrix} 0 & 1/4 & 1/4 & 1/2 \\ 1 & 0 & 0 & 0 \\ 0 & 1/2 & 1/2 & 0 \\ 0 & 1/4 & 1/4 & 1/2 \end{pmatrix},$$

$$S = \begin{pmatrix} 1/4 & 1/4 & 1/4 & 1/4 \\ 1/2 & 1/4 & 1/4 & 0 \\ 0 & 1/4 & 1/4 & 1/2 \\ 1/4 & 1/4 & 1/4 & 1/4 \end{pmatrix}.$$

In this case all B, C combinations have probability 1/16.

We can now find the entropy of the field by considering the conditional probability distribution of D given A, B, C. Clearly $C = L$ gives zero contribution to the entropy. Because of the choice $d = 0$, a W symbol cannot be followed by a U. Thus the next symbol is either given as W because the one above was U, or it is not yet covered, in which case L is the only possibility. Thus the entropy is also zero in this case. The two remaining cases, R and U, are identical. If $B = R$, D can be chosen as either L or U, and we get an entropy of 1 bit. For $B = U$, the only choices are U or W, depending on the symbol above, and the entropy is zero. For $B = L$ or W there are two situations: the following symbol is forced to be W with probability 1/4 ($A = U$), or there are two choices for both A and D. In this case we can use iterative scaling, (7.16) and (7.17), or solve directly to get the marginals 1/4 and 1/2. Clearly the conditional probability distribution is $(1/3, 2/3)$ in both cases.

Thus for these parameters we get the entropy

$$H = 1/8 + (3/16)H(1/3) = 0.297.$$

With a more general version of S given by

$$S = \begin{pmatrix} b+c-e-f & e & f & 1-b-c \\ 2e+2f-c & b+c-e-f & 1-b-e-f & 0 \\ 0 & 1-b-c & b & c \\ 1-b-e-f & f & e & b \end{pmatrix},$$

we get three conditions on the variables by calculating the matrix products. These can be satisfied by selecting c and e, eliminating b and d as $b = 1 - 1.5c + (c^2 - f)/(2e + 2f)$ and $d = (c - e - f + ce + cf - c^2)/(c - e - 2f)$, and finally determining f as the solution of $-(1 - c)^2 + b + d(1 - c - e - b + f) = 0$. Now all admissible configurations have nonzero probabilities, and, by varying the two parameters, we can get an entropy value greater than 0.35 for $c = 0.535$ and $e = 0.305$.

7.9   Notes

The general PRFs including the case with different horizontal and vertical Markov chains were introduced in [1]. Later the issue of defining stationary random fields based on the distribution of 2 by 2 elements was treated [2] and a complete characterization in the binary case revealed alternatives, albeit maybe less natural to the PRF. Coding for the hard-square constraint was treated in detail in [3], where a model shifting the second row of the two-row Markov chain one step to the left is given. The cylinder approach using hidden states in order to define stationary solutions was introduced in [4].

Exercises

7.1 Bands with minimum distance between 1s.
   (a) In a rectangular grid, consider the constraint that each 1 is surrounded by 12 0s at distance 1 and 2 (one norm). Find a finite-state description of a band of three rows satisfying this constraint.
   (b) Find the maximal entropy of the band, and give an upper bound on the entropy of the 2-D process.
   (c) In a hexagonal grid, consider the constraint that each 1 is surrounded by six 0s at distance 1. Find a finite-state description of a band of three rows satisfying this constraint.
   (d) Find the maximal entropy of the band, and give an upper bound on the entropy of the 2-D process.
7.2 Rectangles with three colors.
   (a) Let the Markov chain with three symbols have a transition matrix of the form

$$\begin{pmatrix} 1-2q & q & q \\ q & 1-2q & q \\ q & q & 1-2q \end{pmatrix},$$

   where $q > 1/3$. What is the stationary distribution? Find the reverse chain.
   (b) Find the distribution of three consecutive symbols.
   (c) Choose a distribution for 2 by 2 symbols.
   (d) Calculate the entropy of the field.
   (e) Write a program that outputs N by N samples from the field.
7.3 No isolated bits (n.i.b.) as a PRF. The notation for the transition matrices is as in Section 7.8.2. As in Example 7.12, we take all states equally likely ($s = 0.25$), but use the condition $a - d - f + g = 0$.
   (a) With the parameters $a = 0.2$, $d = 0.3$, and $b = 0.2$, find the transition matrices **R** and **S**.
   (b) Use iterative scaling to find (some of) the conditional probabilities for the PRF.
   (c) With the parameters of the example, find $P[C = 01, B = 10]$. For these values of B and C, find $P[A|BC]$ and $P[D|BC]$.
   (d) Show that a possible solution for $P[AD|BC]$ in this case is the product $P[A|BC]P[D|BC]$.
   (e) Argue that this solution gives the maximum entropy.
7.4 Consider a binary field with the constraint that any 2 by 2 square can contain at most one 1.

(a) Can the constraint be satisfied for a Pickard field (with single bits as variables)?

(b) Find a transition matrix $\mathbf{R}$ for a Markov chain of 1 by 2 blocks such that the constraint is satisfied.

(c) Find a transition matrix $\mathbf{S}$ for the other diagonal by reflecting the states in a vertical axis.

(d) Give a condition for $\mathbf{R}$ and $\mathbf{S}$ to commute.

(e) Find a distribution for the four variables A, B, C, and D such that the constraint is satisfied.

(f) Find the conditional probabilities for D.

(g) Find the entropy of the field.

7.5 Consider defining a PRF for the hard-square constraint using a horizontal boundary and a diagonal boundary specified by

$$\begin{pmatrix} - & A & B \\ C & D & - \end{pmatrix}.$$

(a) Specify two Markov chains defining the boundary. What are the parameters and constraint for these two Markov chains?

(b) Determine $P(D|ABC)$ consistently with the boundaries to define a PRF for specific parameter values of the two Markov chains, possibly using iterative scaling.

(c) Calculate the entropy, $H(D|ABC)$.

(d) Vary the parameters to increase the entropy, $H(D|ABC)$.

(e) Determine the maximal value of $H(D|ABC)$ for the hard-square constraint.

(f) Generate one or more images using the model and estimate the entropy on the basis of these images. Compare your result with the value of $H(D|ABC)$ you calculated.

7.6 The bit-stuffing technique introduced above is a means of mapping an information sequence into a 2-D constrained codeword. It may readily be used for constraints that do not introduce conflicts, e.g. constraints such that it is always possible to write a 0.

(a) For the minimum-distance constraint in which each 1 is surrounded by eight 0s,

$$\begin{array}{ccc} 0 & 0 & 0 \\ 0 & 1 & 0 \\ 0 & 0 & 0 \end{array}$$

generate a K by K square using bit-stuffing with a biased probability toward writing a 1 whenever possible.

(b) Estimate the entropy on the basis of one or more images generated using bit-stuffing of the constraint.

(c) Search for the parameter(s) optimizing the entropy.

(d) Compare your result with the bound(s) obtained using one or more of the bounds defined.

References

[1] D. K. Pickard, "Unilateral Markov fields," *Adv. Appl. Prob.*, **12** (1980), 655–671.

[2] F. Champagnat, J. Idier, and Y. Goussard, "Stationary Markov random fields on a finite rectangular lattice," *IEEE Trans. Inform. Theory*, **44** (1998), 2901–2916.

[3] R. M. Roth, P. H. Siegel, and J. K. Wolf, "Efficient coding scheme for the hard-square model," *IEEE Trans. Inform. Theory*, **47** (2001), 1166–1176.

[4] S. Forchhammer and J. Justesen, "Entropy bounds for constrained two-dimensional random fields," *IEEE Trans. Inform. Theory*, **45** (1999), 118–127.

8 Reed–Solomon codes in applications

8.1 Introduction

In Chapter 3 we introduced linear error-correcting codes, and discussed how long codes can be obtained from short codes by the product construction. Reed–Solomon (RS) codes over larger alphabets were presented in Chapter 4, where we discussed the Datamatrix format as an example of how RS codes can be used to protect binary data. In this chapter we continue the analysis of these themes insofar as they relate to error correction in video and other applications.

We describe constructions of long error-correcting codes that are suitable for encoding of two-dimensional (2-D) media. The size of the pages makes it desirable to have relatively long codes, and 2-D constructions are often used to obtain long codes. However, there is not necessarily a link between the 2-D structure of the media and the code. In the last section we suggest that very long codes could be given a structure that would allow the 2-D structures to be connected, and that such a code could be partially decoded in cases in which only a subset of the data has to be retrieved.

8.2 Binary images of Reed–Solomon codes

The RS codes that are used in applications are always based on the fields $F(2^m)$. Here $m = 8$ is the traditional choice, but future applications are likely to use larger fields. As discussed in Section 4.6, the field is often constructed from a so-called primitive polynomial, $p(z)$. The elements are represented as m-bit vectors (which are interpreted as coefficients of polynomials of degree less than m), and in particular $\alpha = (0100\ldots)$ (representing the polynomial z) is a primitive element.

We noted in Section 4.5 that the codewords of an (N, K) RS code are multiples of the *generator polynomial*

$$g(x) = (x - \alpha)(x - \alpha^2)\ldots(x - \alpha^{N-K}).$$

Here the codewords are interpreted as polynomials over $F(q)$,

$$f_{N-1}x^{N-1} + f_{N-2}x^{N-2} + \cdots + f_1 x + f_0,$$

where the coefficients f_j are elements from $F(2^m)$ (the variable x should not be confused with z that is used in the field elements).

Reed–Solomon codes can be converted into (linear) binary codes in several ways. A direct and important way is to represent the elements of the field $F(2^m)$ as an m-bit binary vector.

The *binary image* of the (N, K) RS code is a linear binary (Nm, Km) code. In any codeword of the original RS code, each element β of $F(2^m)$ is replaced by a binary m by m matrix, where the rows are the binary representations of

$$\beta, \alpha\beta, \ldots, \alpha^{m-1}\beta.$$

Since the original codeword is scaled by elements of $F(2^m)$, the corresponding word in the binary image is found as a linear combination of these m binary rows. Similarly, if the generator matrix of the RS code is given as a K by N matrix with elements from $F(2^m)$, the generator matrix of the image is obtained by replacing each element by the corresponding binary matrix.

If the minimum distance of the original code is $D = N - K + 1$, the minimum distance of the binary image is at least D. In many cases there are codes of length Nm with larger minimum distances, but we cannot improve the bound on the minimum distance of the image without making additional assumptions about the way in which the field elements are represented. The exact minimum distances of the codes used in applications are not necessarily known.

The Datamatrix code in Example 4.8 is an example of a binary image of an RS code over $F(256)$. Most current applications use codes defined over the field $F(2^8)$, and the length is limited to 255 bytes or 2040 bits. The code length is often reduced to fit the frame size by shortening, i.e. the first information symbols are set to zeros and eliminated. If the frame size exceeds 255 bytes, it can be segmented into several codewords. This is the usual approach in image files.

However, it is often advantageous to use *interleaving* of several words. A frame of codewords $(c_{i1}, c_{i2}, \ldots, c_{iN})$ of length N and interleaving degree I consists of NI symbols,

$$(c_{11}, c_{21}, \ldots, c_{I1}, c_{12}, c_{22}, \ldots, c_{I2}, \ldots, c_{1N}, c_{2N}, \ldots, c_{IN}).$$

Interleaving facilitates parallel processing, and in addition it makes the code more robust against errors that occur in clusters. In the Datamatrix standard, interleaving is specified for large labels. For communication on optical fibers, interleaved RS codes are used in the Synchronous Digital Hierarchy (SDH or SONET). Here the interleaving is consistent with the multiplexing of the data streams, and the codes can be decoded in separate streams. When protection against clustered errors is part of the purpose of interleaving, it should take place at the symbol level. Bit interleaving would distribute a packet of errors into more RS symbols.

Example 8.1 (Blu-ray disk format). *The Blu-ray format is used for video disks with larger storage capacity than had hitherto been available. On the track the bits are limited to run lengths between 2 and 8 for both types, and there is a further limitation on the running sum as discussed in Chapter 2. The data format can be described as*

496 by 152 arrays of octets. The data are entered along rows, and, due to the physical
properties of the disk, bursts of errors can occur in this direction. The data are protected
by (248, 216, 33) RS codes over $F(256)$ with two interleaved codewords in each column
(referred to as the long-distance code). In addition, four columns (referred to as pickets)
are inserted (spaced 38 columns apart) encoded with (62, 30, 33) RS codes, eight in
each column. When errors are corrected in the same row and two consecutive pickets,
it is assumed that a burst has occurred, and the data between the pickets are erased.
The erasures and any random errors in the data are corrected by the long-distance
code.

8.3   Concatenated codes

The minimum distance of the binary image is at least D, the distance of the RS code,
and in general it is difficult to prove that the distance is larger. However, codes with
larger minimum distances can easily be constructed. Each symbol may be encoded as a
codeword of an (n, m, d) binary code to obtain an (nN, mK) linear code, traditionally
referred to as a concatenated code.

The minimum distance of the concatenated code is at least Dd, since the RS codeword
has at least D nonzero symbols, each of which is represented as a binary vector of weight
at least d.

Larger blocks can be constructed by interleaving several RS codes. We can think of a
codeword as an array with I rows, a RS codeword in each row. A concatenated code can
then be constructed by encoding each column as an (n, Im, d) linear code, and since n
and k are now larger than for a single code, we can also get a larger value of d or in
general a more efficient code.

Example 8.2 (Concatenated codes).   *Some good short binary codes can be constructed*
as concatenated codes. Let the RS code be (16, 10, 7) over $F(16)$ and use a single parity
check, (5, 4, 2), as the inner code. The result is an (80, 40, 14) binary code. Several
applications use (255, 223, 33) or (255, 239, 17) RS codes with interleaving degree 5
as outer codes and binary inner codes of rate 1/2. With the (80, 40, 14) code we would
get minimum distances of at least 462 and 238 (actually a binary code with minimum
distance 10 and a different structure is used as inner code in these standards).

The term "concatenated" refers to systems in which encoding and decoding of the
"inner" binary and "outer" RS codes take place in separate stages. In more complex
media or communication systems the two-level concept offers great flexibility which
can be combined with excellent performance.

The processing of the inner, "physical," layer can incorporate run-length constraints,
properties of the modulation format in a transmission system, and possibly synchronization issues, in addition to the correction of most of the errors. The decoded symbols

obtained in this way can then be passed through the RS decoder, which will correct any residual errors. The symbol error probability from the inner system will usually have to be measured on the actual system or found by simulation of a suitable model. Once the error probability is known, the performance of the outer stage can be calculated.

Since the RS decoder corrects any combination of $T = (N - K)/2$ errors and no other error patterns, the probability of decoding failure for independent errors of probability p is given by the binomial distribution

$$P_f = \sum_{t>T} \binom{N}{t} p^t (1-p)^{N-t}.$$

If the errors are not independent, the probability usually increases significantly. Thus interleaving is often used to spread out dependent errors. When $t' > T$ errors occur, the decoder will most often not produce any result, and the failure is detected. In rare cases, T errors are corrected, and the result is an undetected error. The number of correctable error patterns is

$$\sum_{t \leq T} \binom{N}{t} (Q-1)^t \approx Q^{2T}/T!, \tag{8.1}$$

where Q is the size of the alphabet, and $N \approx Q$. The number of distinct syndromes is Q^{2T}, and the probability that a randomly chosen syndrome is decoded is about

$$P_e = 1/T!. \tag{8.2}$$

This quantity is a very good approximation to the probability that an error pattern with more than T errors is decoded to another codeword. Thus decoding errors occur quite frequently for small values of T, but from $T = 8$, which is a commonly used standard value, undetected errors are quite rare.

8.4 Product codes

As discussed in Section 3.3.6, two codes with parameters (n_1, k_1, d_1) and (n_2, k_2, d_2) can be combined to form a product code with parameters $(n_1 n_2, k_1 k_2, d_1 d_2)$. We often think of such a code as a 2-D structure in which codewords are arrays of data and the error-correcting codes are applied to rows and columns. In this chapter we consider products of RS codes in particular.

The use of such codes is sometimes motivated by the nature of the data and the error mechanisms that affect the transmitted words, but they may also be used to get long codes that can be decoded by several applications of simple decoding methods.

Example 8.3 (The DVD code). *For video DVDs a product of two RS codes, namely* (208, 192, 17) *and* (182, 172, 11), *is used as part of an encoding process that also includes data compression and a run-length-constrained binary code. The CD and DVD codes use two different ways of ensuring that two binary 1s are separated by at least two* *and no more than ten zeros. Each 8-bit symbol is converted to 17 bits in the CD code and 16 bits in the DVD code. It was shown in an earlier example that an encoding with distance at least 3 between 1s is possible with a rate of slightly more than 0.5.*

One of the significant advantages of product codes is that simple decoders for the component codes can be combined to decode the total code. For products of RS codes, we can usually assume that there are very few decoding errors (if t is small, we can use a code with even distance, $n - k = 2t + 1$, and get the same effect). Thus, after decoding the columns, some of them are left undecided. If the corresponding symbols are considered unknown (erased), the rows can be decoded by solving a system of linear equations. Thus at least $(t_2 + 1)(t_1 - 1)$ or close to $d_1 d_2 / 2$ errors are always corrected.

When the column decoding fails, we may choose to leave the codeword as received, i.e. with the $t_2 + 1$ or more errors. If these happen to occur in the same rows, we are back to correcting only $d_1 d_2 / 4$ errors. However, if the errors occur in random positions of the columns, the error probability in the rows is much reduced, and we can expect more errors to be corrected in the second stage. If this approach is taken, we can get even better results by repeating the decoding alternating between rows and columns. The performance of such a decoder is discussed in the following section.

8.5   Iterated decoding of product codes

The decoding of the product codes can be better understood by phrasing the problem in graph-theory terms. A graph is composed of a set of nodes and a set of branches, each connecting a pair of distinct vertices. We consider a special type of graph, called bipartite. Here the vertices are divided into two disjoint subsets, the right set and the left set. Each branch connects a right vertex to a left vertex.

A product code can be described by a bipartite graph: each row code is represented by a left vertex and each column code by a right vertex. The code symbols are labels on the branches. In this way the symbols on branches that meet in a left vertex must be a codeword in the corresponding row code and similarly the labels on branches that meet in a right vertex must be a codeword in the column code. Since each column code shares one symbol with each of the row codes, and vice versa, there is exactly one branch connecting each pair of left/right vertices. Such a bipartite graph is called complete. The decoding problem can be described by indicating the position of each error by a special label on the corresponding branch. Actually, we can remove all other branches and just keep the ones that are in error. Decoding of the column codes is now implemented by finding the number of branches (with errors) connected to each right vertex. Whenever there are at most t_1 branches, the errors are corrected, i.e. the branches are removed. Since there are no branches connected to the vertex, we may remove it also. The left side is now processed in the same way to represent decoding of the row codes. We alternate between decoding right and left vertices until there are no further changes. Either the result is an empty graph where all errors have been successfully decoded, or we are left with a subset of right and left vertices such that each right vertex is connected to at least $t_1 + 1$ left vertices and each left vertex to at least $t_2 + 1$ right vertices.

In the case $t_1 = t_2 = t$, such a subgraph is called a $t + 1$ core, and the iterated decoding is a way of determining whether the original error graph had such a core. An important result in graph theory tells us when such a core is likely to exist in a random graph, and we use this result to get an estimate of the error-correcting capability of a product code.

If the errors occur at random positions in the array with some probability p, the average number of errors in each code is pn and that in the total array is pn^2. Clearly, decoding is not possible if pn is much larger than t, so assume for the time being that $pn = t + 1/2$. Clearly about half of the columns contain more than t errors and the other half at most t errors. Thus a little less than half of the errors are corrected by the row codes. The remaining errors are almost randomly distributed in the columns, and, since the density is reduced, a significant fraction of them will be corrected. As the process is repeated it becomes difficult to analyze the distribution of the errors and predict whether all errors are eventually decoded. However, there is a result in the theory of random graphs that gives the answer: if the total number of errors is less than $nc_{t+1} - \epsilon$, there is a very small probability of a $t + 1$ core (going to zero for fixed t and n going to infinity), whereas if the number of errors is greater than $nc_{t+1} + \epsilon$, there is a high probability of a large $t + 1$ core (a large fraction of the codes will not be decoded). Here the threshold can be expressed in terms of the Poisson distribution:

$$c_k = \min_{\lambda}[\lambda/\pi_k(\lambda)], \quad \lambda > 0, \pi_k(\lambda) = P[\text{Poisson}(\lambda) \geq k - 1]. \tag{8.3}$$

Thus $c_3 = 3.35$, $c_4 = 5.14$, $c_5 = 6.80$, $c_6 = 8.37$, and $c_9 = 12.78$. Asymptotically $c_k \approx k + \sqrt{k \log k}$.

The result shows that the product code has a remarkable performance: even when more than t errors occur on average in each row and column, iterated decoding will eventually with high probability correct all the errors.

Example 8.4 (Iterated decoding). *For $T = 8$ the threshold is found from (8.3) to be more than 12 for sufficiently long codes. For a product of two $(255, 239, 17)$ RS codes, there is a high probability of decoding 2800 errors. The minimum distance of the code is found from (3.9) to be 289. The result is verified by an easy computer exercise. After introducing about 4.5% 1s into a 255 by 255 matrix, set a row or column to zero whenever the weight is less than 9, and follow the reduction of the total weight of the matrix. With a slightly larger fraction of 1s, the process fails.*

The decoding of products of two RS codes with different error-correcting capabilities, t_1 and t_2, can be analyzed in the same way, but there is no simple expression for the number of errors corrected.

Example 8.5 (Iterated decoding of the DVD code). *The two component codes correct 8 and 5 errors. These codes were chosen to give a good performance with the actual distribution of errors and only one round of decoding for each code. However, if a* *random distribution of symbol errors is assumed and the decoding is repeated, in total 1950 errors are decoded in each frame with high probability. Thus, when the (208, 192) codes in the columns are decoded first, there can be an average of 9.37 errors in each word. After the decoding, the average number of errors in the rows is 9.0, and many of these errors are corrected by the (182, 172) code. All errors are typically corrected within five iterations.*

Example 8.6 (A modified product code for optical transmission). *As mentioned in Section 8.2, the basic standard for synchronous optical transmission (ITU G-975) specifies interleaved binary images on the basis of (255, 239, 17) RS codes. However, several more powerful systems have been proposed. One such code can be described as a modified product or concatenated code. A codeword is an array in which the rows are encoded by 16 RS codes over $F(2^{10})$ and the columns by 64 binary codes (so-called BCH, which we do not discuss, but they are decoded by the same basic algorithm as RS codes). Thus each component code covers several rows or columns in the array. Both component codes correct eight errors. Both codes are shortened to fit the actual frame size, but we shall modify the numbers slightly to simplify the discussion. Let the RS codes be (768, 752) and the binary codes (2015, 1920). Thus the 16 times 7680 bits in the RS codes are encoded as the 64 times 1920 information symbols of the binary codes. A single RS code shares 120 bits with each binary code, and for low error probabilities we can estimate the probability of decoding failure as the probability that any such set of bits contains nine errors:*

$$P_f \leq 16 \times 64 \binom{120}{9} p^9.$$

If we assume that iterated decoding of the component codes corrects other error patterns, and that in the case of decoding failure other errors in the frame are decoded, there are nine errors left among the $64 \times 16 \times 120$ bits, which lets the system reach the target error probability of 10^{-15} for $p \leq 10^{-3}$.

8.6　Graph codes for two-dimensional error-correction

The analysis of product codes can be generalized to larger code structures based on bipartite graphs. For simplicity we consider only the case in which the graphs are regular, i.e. all vertices have n edges, and this is also the length of the component codes. Each edge carries a code symbol, and, if the number of vertices on each side is s, the total length of the code is $N = ns$. The symbols associated with edges that meet in a vertex must be a codeword in the component (n, k, d) code. For a given component code, we may choose s significantly larger than n and get a code that is much longer than the product code.

Example 8.7 (Cross-interleaved RS codes in the CD format). *In the original CD format, 24 bytes of data are encoded using a shortened* $(28, 24, 5)$ *RS code over* $F(256)$. *The choice of this code is related to the limited technical possibilities at the time the standard was developed. These 28 data streams are then subjected to different delays, and subsequently encoded again using* $(32, 28, 5)$ *RS codes. The resulting data are again interleaved to allow very long damaged segments of the recording track to be reconstructed. This form of interleaving can be described in a simplified way by a graph where s nodes on the right represent encoding of the* $(28, 24)$ *code, and node j on the right is connected to nodes* $j + d_1, j + d_2, \ldots, j + d_{28}$ (mod s) *on the left. The left node represents the encoding of a* $(32, 28)$ *code and would thus add four additional parity symbols.*

Since the number of parity checks in the total code is $2s(n - k)$, the dimension of the total code is at least $K \geq 2sk - sn$ (some checks may be linearly dependent).

The properties that could make such structures attractive for future application are that very large page sizes can be encoded and decoding is still based on short component codes with low decoding complexity. Good performance is possible, and partial decoding of a page is possible if the user wants to access just a particular segment.

8.6.1 Bipartite graphs from geometries

To get a code with good error-correcting properties, it is desirable that the graph is well connected. The following two criteria can be used to measure the quality of the graph.

Let the distance between two vertices be measured as the length of the shortest path from one to the other (expressed as the number of edges). The diameter of the graph is then the largest distance between two points. For given n and s, we want the diameter to be small.

A circuit is a path of distinct edges from a vertex back to itself. The girth of the graph is defined as the length of the shortest circuit. A good graph for our purpose should have a large girth.

A closely related property (which we do not define precisely) is called the expansion of the graph. A good expander has the property that for any small subset of the nodes, Δ, the subset that is within some distance of Δ is much larger. By constructing a code from a graph with such properties we ensure that errors in some part of the total codeword can be corrected by involving symbols in other parts of the graph.

Good bipartite graphs can be obtained from geometries over finite fields. In this chapter we use a simple example that has familiar properties. A finite Euclidean plane over the finite field $F(Q)$ consists of points (x, y) with coordinates in $F(Q)$ and lines $y = ax + b$, where the coefficients (a, b) are constant in $F(Q)$. Thus there are Q^2 points and lines, since we have left out lines of the type $x = c$ to get a construction that is symmetric in the two sets of objects. We can now define a bipartite graph by labelling the Q^2 vertices on the left by the points and the vertices on the right by the lines. A point vertex (x, y) and a line vertex (a, b) are connected by an edge whenever $y = ax + b$.

Thus all vertices have q edges. Starting from any particular vertex, say on the right, Q vertices on the left are reached by a single edge. By going one step further, we reach $Q(Q-1)$ new vertices on the right. Thus almost all vertices on this side are reached, and only the original vertex is reached by more than one path. In the next step, all vertices on the left are reached close to Q times. The graph has diameter 3 and girth 6.

By using (Q, k, d) RS codes as component codes, we get

$$N = Q^3, \tag{8.4}$$

$$K \geq Q^2(2K - Q). \tag{8.5}$$

8.6.2   Minimum distances of graph codes

It follows from the description in Section 8.6.1 that, if the minimum distance of the component code is d, no codeword in the total code can have weight less than $d^3 - 2d^2 + 2d$. Starting from a nonzero vertex, there are at least d edges with nonzero labels, each connected to at least $d(d-1)$ new nonzero edges, etc. The codeword cannot be complete unless all edges leading into a node are zero or at least d of them are nonzero.

However, for a large graph with good expansion properties there is a stronger bound. Let the edges in the bipartite graph be indicated by 1s in the $2s$ by $2s$ matrix

$$A = \begin{bmatrix} 0 & M \\ M' & 0 \end{bmatrix}.$$

It follows from the regularity of the graph that the largest eigenvalue is n and that the corresponding eigenvector is the all-1s vector. A good expander can be characterized by the property that the second-largest eigenvalue is of the order $2\sqrt{n}$, and for the graph considered here it is $\sqrt{Q}$. We can bound the number of nodes corresponding to nonzero codewords by considering a vector that is 1 in those α positions and $-b$ elsewhere. To be an eigenvector for the second eigenvalue, it must be orthogonal to the all-1s vector,

$$\alpha - (s - \alpha)b = 0.$$

Furthermore, the nonzero nodes on one side must be connected to at least d nonzero nodes on the other side of the graph. We can now express the product of A and the eigenvector as

$$d - (n - d)\alpha/(s - \alpha) = \sqrt{Q}$$

for the nonzero nodes. From this relation we find the number of nonzero nodes as

$$\alpha = s(d - \sqrt{Q})/(n - \sqrt{Q}).$$

Thus the minimum weight of the code is lower bounded by

$$D \geq ds\frac{d - \sqrt{Q}}{n - \sqrt{Q}}. \tag{8.6}$$

Clearly the bound is useful only for rather large n for which the rate k/n can be at least moderately high while $d > \sqrt{n}$.

Iterated decoding of the component codes can be analyzed as discussed for product codes. In particular, if symbols are left unchanged when a component code is not decoded, and $d/2$ is large enough that we can neglect the probability of decoding error, the estimate of the correctable fraction of errors is still given by (8.3).

8.6.3 Graph codes as two-dimensional arrays

The codes described in this section can be decoded by many applications of the decoding algorithm for the component codes, leading to low complexity and the possibility of parallel processing. However, it is important that the symbols can be conveniently accessed for decoding each component code. To facilitate the storage access and as a step toward applications in 2-D media, we arrange the symbols in a quadratic array.

Assuming that Q is a square, we want a codeword to be a $Q\sqrt{Q}$ quadratic matrix, where each row and column consists of $\sqrt{Q}$ codewords from the component code. The interleaving of these codewords has to vary from row to row in order to get the favorable distance of the graph code, but the processing is still facilitated to a significant degree.

The first step in this reorganization of the code is to collect the bundles of Q parallel lines $y = ax + b$ which have a particular value of a. These codes are associated with $\sqrt{Q}$ rows in the matrix. Similarly, the Q points which have a particular value of x (these points are not on a line since we did not include the lines $x = c$ in the construction) are collected and associated with $\sqrt{Q}$ columns. A bundle of lines and points will intersect in Q parameter combinations, giving a symbol in each of the Q codes. Thus we have achieved part of the goal by restricting each component code to $\sqrt{Q}$ rows or columns.

We can go on to separate the symbols in such a way that the component codes are constrained to be located in a single row or column. We write each symbol in $F(Q)$ as a pair over the smaller field of $\sqrt{Q}$ symbols. With this notation we can place points with the same first component in the expansion of y in a single column, and similarly those lines where the values of b share the second component are placed in the same row. The intersection of a row and a column now contains exactly one symbol.

8.7 Errors in two-dimensional media

The performance of error-correcting codes is usually described in terms of the failure probability or the output bit error probability when errors are assumed to occur independently with a given probability. However, the use of RS codes, which correct symbol errors rather than bit errors, combined with the choice of interleaved schemes and product codes is often related to applications in which errors occur in clusters.

8.7.1 Defects in physical media or in the writing/printing process

Both magnetic and optical media suffer from some permanent errors due to defects in the magnetic/reflecting coating or similar defects associated with the production or writing processes. Such errors may extend over several bits, but the extent is typically of the same order of magnitude. Thus, in a track or line, clusters of bit errors may occur, but 2-D spots or errors in nearby tracks are usually found only in actual 2-D formats. Irrespective of whether the bits are used directly in an RS code or first processed by a decoder for constrained sequences/fields, the error is likely to affect only a single RS symbol. Thus the large symbol alphabet of the code provides the required robustness in such cases. It may be noted that small permanent errors can have a significant effect on headers and synchronizing patterns that are used to locate the beginning of codewords. Thus the unprotected header information on CDs containing frame numbers and timing information may be corrupted. The long RS codes place an increased demand on the synchronization system.

8.7.2   Degradation of the media in use

Storage media are subject to degradation from being handled and read. Some physical processes, such as exposure to light or magnetic fields, may cause general damage, but mechanical handling is more likely to leave scratches that affect only a small part of the data, but an area extending over a large number of bits. It is impractical to correct such an error by use of a single code, and the common approach is to use interleaving as a way of distributing the errors over many different codewords. If the symbols of a product code are properly organized, it can function in a similar way.

In the original CD format, extensive interleaving is used to correct surface errors by application of shortened two-error-correcting RS codes. However, significant additional robustness is obtained by using longer multiple-error-correcting codes as in the DVD format.

If segments of the medium are permanently damaged and the encoder has knowledge of the locations of these segments (from checking the stored information), it is possible to keep a record of the segments and work around them. Such methods are useful on bulk storage media such as hard disks, but will not be discussed here.

8.7.3   Degradation due to loss of alignment or synchronization

Reed–Solomon codes rely heavily on correct synchronization, since each codeword is long and the proper segmentation of a string of symbols cannot be derived in a practical way from the structure of the codewords. In 2-D media there are additional problems relating to the alignment of the page and possible geometric distortion, either of the physical medium or during the reading process. Such problems have to be corrected on a lower level, by the use of markers, constrained codes, or other properties of the bit-level code. In a concatenated code, the inner code may include constraints for this purpose.

8.8   Notes

Some of the standards for communication are issued by the International Telecommunication Union (ITU), the European Telecommunications Standards Institute (ETSI), the

European Broadcasting Union (EBU), or the International Organization for Standardization (ISO). Such standards may be available online from the organizations, but in some cases they have to be purchased. Standards for consumer products are often available online from groups of manufacturers who have developed them.

Exercises

8.1 What are the parameters of the binary image of a $(31, 20)$ RS code over $F(32)$? Add a parity check to each symbol (use a $(6, 5, 2)$ inner code). What are the parameters of this concatenated code?

8.2 Consider the same $(31, 20)$ RS code, but use interleaving degree 2. What are the parameters with a single parity check on each column? Use a $(15, 10, 4)$ binary code as the inner code. Find the parameters of the resulting code.

8.3 Consider the product of two $(63, 59)$ RS codes over $F(64)$. What is the rate of the code and the length in bits? Find a binary image of an RS code over $F(2^{12})$ with approximately the same length and rate. Compare the error-correcting capabilities of the two codes.

8.4 Assume that the probability of symbol error in a $(255, 239)$ RS code over $F(256)$ is 0.02. What is the probability of decoding failure? What is the probability that a received word is decoded to a word different from the one transmitted?

8.5 Write a program that simulates the decoding of a product of two $(31, 23)$ RS codes. Is it acceptable to assume that no decoding errors occur when there are more than T errors? Modify the simulation to take decoding errors into account.

8.6 A graph code is constructed from a Euclidean plane graph and component $(16, 12, 5)$ RS codes. What is the length of the code? Give a lower bound on the dimension and the minimum distance. Convert the graph code into a square array. What is the size of the array, and how many codes are interleaved in each row/column?

Appendix A Fast arithmetic coding

A1 Fast binary arithmetic coding

In many practical applications a fast version of arithmetic coding is desired. In the JBIG standards mentioned, as well as other image- and video-coding standards, multiplications and divisions both of the arithmetic coding and of the conditional probabilities are avoided. The example below presents an approximative solution that avoids multiplications and thereby speeds up the basic recursions.

A1.1 Multiplication-free arithmetic coding

Consider a binary arithmetic coder and let q denote the smallest of the two probabilities, i.e. $\min\{P(x_{n+1} = 0|x^n), P(x_{n+1} = 1|x^n)\}$. Let LPS denote this least probable symbol and MPS the most probable symbol. The interval shall be represented by the lower bound on the interval, l_n, corresponding to (the codeword) $C(x^n)$ and the interval width $A(x^n)$. In each step, the interval is scaled by powers of 2 such that $3/4 \leq A(x^n) < 3/2$. This scaling is easily implemented by left-shifting the registers. By approximating $A(x^n)q \approx q$ the recursion is simplified. On placing the MPS in the lower part of the interval, the multiplication-free recursion becomes

$$A_n = \begin{cases} A_{n-1} - q & \text{for} \quad x_n \text{ MPS}, \\ q & \text{for} \quad x_n \text{ LPS} \end{cases}$$

for the interval width, and l_n is given by

$$l_n = \begin{cases} l_{n-1} & \text{for} \quad x_n \text{ MPS}, \\ l_{n-1} + A_{n-1} - q & \text{for} \quad x_n \text{ LPS}. \end{cases}$$

The approximation eliminates the multiplication and the precision of the LPS probability q is directly related to the precision. Again the carry problem may be solved by bit-stuffing. The loss in coding efficiency is in practice fairly small.

The probabilities assigned for coding are based on the adaptive estimate given by $p_0 = (n_0 + \delta)/(n_0 + n_1 + 2\delta)$ and the occurrence counts (n_0, n_1) in each context. To avoid the need for division, an approximate calculation (and update) is implemented by a finite-state machine for speed, thus also eliminating the division for the basic probability estimate. To avoid having states for all values of (n_0, n_1), which would be the case in a look-up-table solution, the states are updated on the basis of a probabilistic scheme, e.g. each time a new coded bit is defined. Thus, each time an LPS occurs, the estimate is updated. When an MPS leads to an update, the increment reflects the number of MPSs since the last MPS update in the given context based on the expected value.

Appendix B   Maximizing entropy

B1   Introduction

Some problems involving maximizing the entropy of memoryless sources and Markov sources can be solved by a classical approach called *Lagrange multipliers*. This appendix gives a brief introduction to the technique, and develops two important results for Markov sources.

B2   Maximum-entropy memoryless sources

If the probability distributions for a source are not known in detail, it is often useful to construct a model that has maximum entropy. For memoryless sources with a finite alphabet, the entropy is maximized by letting all symbols have the same probability. However, when the alphabet is large, or even infinite, the distribution may have to satisfy one or more constraints. We consider the particular cases in which the alphabet is the integers, the positive integers, and a finite subset of the integers. We can then find corresponding maximum-entropy distributions that have given values of the mean, the variance, or other functions of the distribution.

Let the probability of the source symbol i be p_i, and let a constraint on the distribution have the form

$$\sum_i b_i p_i = B.$$

We then add the constraint to the function we want to maximize multiplied by a parameter, μ, the *Lagrange multiplier*. The constraint that the p_i sum to unity is treated similarly, and we proceed to find the optimum by taking partial derivatives:

$$\frac{\partial}{\partial p_i} \left[\sum_i (-p_i \log p_i) + \mu \sum_i b_i p_i + \nu \sum_i p_i \right] = -\frac{1}{\ln 2} - \log p_i + \mu b_i + \nu.$$

Consequently the distribution has the form

$$p_i = ca^{b_i}, \tag{B.1}$$

and we can find the values of the two parameters such that the constraints are satisfied.

Example B.1 (A given mean value). *If the alphabet is the positive integers and the mean value is b, we find the maximum-entropy distribution from (B.1) as $a = 1 - 1/b$ and $c = 1/(b-1)$.*

B3 Markov sources with a given stationary distribution

As discussed in Chapter 2, we can express the structure of a finite-state source by the adjacency matrix, **T**, which indicates the possible transitions between two states. For these transitions, we want to choose nonzero transition probabilities p_{ij} to maximize the entropy of the source subject to the constraint that the stationary distribution on the states is the vector $\mathbf{p}^*$ (which may be chosen to match observations). The condition $\mathbf{P}\mathbf{p}^* = \mathbf{p}^*$ is written as a constraint for each row of **T**, while the columns must sum to 1. On taking partial derivatives, we get

$$\frac{\partial}{\partial p_{ij}} \left[\sum_j p_j^* \sum_i (-p_{ij} \log p_{ij}) + \sum_j \mu_j \left(\sum_i p_i^* p_{ij} - p_j^* \right) + \sum_i \nu_i \sum_j p_{ij} \right]$$

$$= -p_j^*(1/\ln 2 + \log p_{ij}) + \mu_j p_i^* + \nu_i.$$

Thus the transition probabilities have the form

$$p_{ij} = a_i b_j. \tag{B.2}$$

The transition matrix is obtained from the adjacency matrix by suitable scalings of rows and columns. The factors can be found by solving a system of equations, but, as discussed in Chapter 7, we usually prefer to apply iterative scaling.

B4 Markov sources with a given structure

As in the previous section, the structure of a finite-state source is given by the adjacency matrix, **T**. Here we want to choose nonzero transition probabilities p_{ij} to maximize the entropy of the source, and the stationary distribution is now also variable.

Initially we do the same calculation as in the previous section and find that the transition probabilities are of the form (B.2). Since the probability of a string of symbols from the source is

$$p_{i_1}^* p_{i_1 i_2} p_{i_2 i_3} \cdots$$

and we get the maximum entropy when the probability is approximately λ^{-n} for a string of length n, we are looking for a solution of the form

$$p_{ij} = a_j/(a_i \lambda).$$

The condition for the stationary distribution gives

$$\sum_i p_i^* p_{ij} = (a_j/\lambda) \sum_i p_i^*/a_i = p_j^*,$$

which shows that p_i^*/a_i is the left eigenvector of the adjacency matrix with eigenvalue λ. We can get the stationary distribution p_i^* by scaling $a_i b_i$, where b_i is the right eigenvector of $\mathbf{T}$ with the largest eigenvalue. Both eigenvectors are easily found by repeatedly multiplying a vector by $\mathbf{T}$ and scaling the result.

Example B.2 (The entropy of a constrained sequence).   *Let the source generate binary triples with even parity. We can describe the structure of the source by the adjacency matrix*

$$\mathbf{T} = \begin{bmatrix} 0 & 0 & 0 & 1 & 1 \\ 1 & 0 & 0 & 0 & 0 \\ 1 & 0 & 0 & 0 & 0 \\ 0 & 1 & 1 & 0 & 0 \\ 0 & 1 & 1 & 0 & 0 \end{bmatrix}.$$

The largest eigenvalue is $2^{2/3}$, and obviously the entropy is 2/3. The left eigenvector is $[1, \lambda/2, \lambda/2, \lambda^2/4, \lambda^2/4]$, and from this vector we find the transition probabilities in the first three columns to be 1/2.

Appendix C   Decoding of Reed–Solomon code in *F*(16)

```
%The Matlab program follows Sections 4.6 and 4.7.
%%%%%%%%%%%%%%%%%%%%%%%%%%%%%%%%%%%%%%%%%
%Generate tables for field calculations
%%%%%%%%%%%%%%%%%%%%%%%%%%%%%%%%%%%%%%%%%
q=16;%size of field
px=19;%irreducible polynomial x^4+x+1
qex(1)=1;%start table of exponents
for ii=1:q-2,
    qex(ii+1)=2*qex(ii);%shift
    if qex(ii+1)>15,%degree 4
        qex(ii+1)=bitxor(qex(ii+1),px);
    end%reduce mod px
end%table of exponents
%%%%%%%%%%%%%%%%%%%%%%%%%%%%%%%%%%%%%%%%%%%
ind=[1:q-1];%indices for tables
%%%%%%%%%%%%%%%%%%%%%%%%%%%%%%%%%%%%%%%%%%%
ql(qex(ind))=ind;%inverse table
%table of 'logarithms'
%%%%%%%%%%%%%%%%%%%%%%%%%%%%%%%%%%%%%%%%%%%
qin=qex(mod(-ql(ind)+1,15)+1);%table of inverses
%%%%%%%%%%%%%%%%%%%%%%%%%%%%%%%%%%%%%%%%%%%
%Nonzero elements can be multiplied by
%adding logarithms modulo q-1
mq=zeros(q); for ii=1:15,
    for jj=1:15,
        mq(ii+1,jj+1)=qex(mod(ql(ii)+ql(jj)-2,q-1)+1);
    end
end%multiplication table, all indices shifted by one
%%%%%%%%%%%%%%%%%%%%%%%%%%%%%%%%%%%%%%%%%%%%

%%%%%%%%%%%%%%%%%%%%%%%%%%%%%%%%%%%%%%%%%%%
%Parameters and generator polynomial
%%%%%%%%%%%%%%%%%%%%%%%%%%%%%%%%%%%%%%%%%%
```

```
n=15;%length of code
k=9;%code dimension
t=floor((n-k)/2);%number of errors corrected
%Find generator polynomial
gx=[1 qex(2)];%alpha a root
for ii=2:6,%add roots
    g1=[gx,0];
    g2=[0,gx];
    for jj=2:ii+1,
        if g2(jj)>0,
            g2(jj)=qex(mod(ql(g2(jj))+ii-1,q-1)+1);
        end
    end
    gx=bitxor(g1,g2);
end
%gx is the generator polynomial

%%%%%%%%%%%%%%%%%%%%%%%%%%%%%%%%%%%%%%%%%%
%Information and systematic encoding
%Input: information sequence, info
%Output: Encoded word, cx
%%%%%%%%%%%%%%%%%%%%%%%%%%%%%%%%%%%%%%%%%%

info=[0 0 0 0 0 0 1 7 9];%information sequence
%info=round(rand(1,k));%random information
cx=[info,zeros(1,n-k)];%start systmatic encoding
for ii=1:k, %perform division by gx
    if cx(ii)>0,
        ff=cx(ii);
        for jj=1:(n-k+1),
            cx(ii+jj-1)=bitxor(cx(ii+jj-1),
            qex(mod(ql(gx(jj))+ql(ff)-2,q-1)+1));
        end
    end
end%division
cx=[info,cx(k+1:n)];%Append the remainder as parity

%%%%%%%%%%%%%%%%%%%%%%%%%%%%%%%%%%%%%%%%%%%
%Add error pattern
%Input: codeword, cx
%Output: received word, rr
%%%%%%%%%%%%%%%%%%%%%%%%%%%%%%%%%%%%%%%%%%%
```

```
ee=rand(1,n);%Generate random errors
pe=.10;%channel symbol error probability
ee=ee<pe;%random error positions
eval=round(rand(1,n)*15);%random values
ee=ee.*eval;%random error pattern
%ee=[0 0 0 0 0 0 7 0 0 12 0 0 0 2];%error pattern
%ee=zeros(1,n)%try no errors
%ee=[1 0 0 0 1 0 0 0 1 0 0 0 0];%another error pattern
rr=bitxor(cx,ee);%received vector

%%%%%%%%%%%%%%%%%%%%%%%%%%%%%%%%%%%%%%%%%%
%Start decoding
%Syndrome calculation
%Input: received word, rr
%Output: syndrome vector, ss
%%%%%%%%%%%%%%%%%%%%%%%%%%%%%%%%%%%%%%%%%%

ss=zeros(1,n-k);%Start syndrome calculation
jii=[1:n-k];%syndrome index
jis=qex(jii+1);%power of alpha
for ii=1:n,
    sn=mq(ss+1,jis+1);
    ss=diag(sn)';
    ss=bitxor(rr(ii),ss);
end%syndrome vector (reversed)

%%%%%%%%%%%%%%%%%%%%%%%%%%%%%%%%%%%%%%%%%%
%Euclid's algorithm
%Input: syndrome vector, ss
%Output: flags for zero errors, zer, too many errors, mer,
%error locator, qv, error evaluator rv,
%expected number of errors, te
%%%%%%%%%%%%%%%%%%%%%%%%%%%%%%%%%%%%%%%%%%

rr1=[zeros(1,n-k),1];%r(-1)
rr0=[ss(n-k:-1:1),0];%reverse syndrome vector
qq1=zeros(1,t+1);
qq0=zeros(1,t+1);%quotients to become error locator
qq0(1)=1;
dgr1=n-k+1;%degree of rr1+1, leading nonzero position
dgq1=0;%degree qq1+1, since 0 at this point
dgr0=find(rr0>0,1,'last');%leading nonzero position (degree+1)
%if no nonzero syndromes, skip Euclid
dd=size(dgr0);%test if nonempty
```

```
if dd(2)>0,
    fac=qin(rr0(dgr0));%inverse of leading coefficient
    rr0=mq(fac+1,rr0+1);%scale to make leading coefficient 1
    dgq0=1;%degree of qq0 is 1
    qq0=mq(fac+1,qq0+1);%scale by the same factor
    flag=1;%flag for stopping condition on degree(r)
    zer=0;%flag for zero errors
else
flag=0;%skip Euclid
zer=1;%zero errors
end
%initial state for Euclid's algorithm
mer=0;%flag for too many errors
if dgr0<t+1,
            flag=0;
            mer=1;
end% The syndrome polynomial has low degree
% from the beginning, skip Euclid
% In this case there are too many errors
phase=1;%rr1 has higher degree
%phase 0 and 1 operate on two sets of variables,
%and their roles are reversed


%%% Initial conditions have been set up

while flag==1,
%do a step of the algorithm until the degree is low
    if phase==1,
        for ii=1:dgr1-dgr0+1,
            tr=mq(rr1(dgr1-ii+1)+1,rr0+1);%scale rr0
            tq=mq(rr1(dgr1-ii+1)+1,qq0+1);%scale qq0 the same way
            tr=[zeros(1,dgr1-dgr0-ii+1),tr(1:n-k+1-dgr1+dgr0),
                zeros(1,ii-1)];%shift rr0
            tq=[zeros(1,dgr1-dgr0-ii+1),tq(1:t+1-dgr1+dgr0),
                zeros(1,ii-1)];%shift qq0
            rr1=bitxor(rr1,tr);%add to rr1
            qq1=bitxor(qq1,tq);%add to qq1
        end%remainder has lower degree
        dgr1=find(rr1>0,1,'last');%find leading term
        fac=qin(rr1(dgr1));%inverse of leading coefficient
        rr1=mq(fac+1,rr1+1);%scale to make leading coefficient 1
        qq1=mq(fac+1,qq1+1);%scale qq1
        %rr1%result after each step
        if dgr1<t+1,
```

```
            flag=0;
        end%end of Euclid
    end
    if phase==0,%rr0 has higher degree
        for ii=1:dgr0-dgr1+1,
            tr=mq(rr0(dgr0-ii+1)+1,rr1+1);%scale rr1
            tq=mq(rr0(dgr0-ii+1)+1,qq1+1);%scale qq1 the same way
            tr=[zeros(1,dgr0-dgr1-ii+1),tr(1:n-k+1-dgr0+dgr1),
                zeros(1,ii-1)];%shift rr1
            tq=[zeros(1,dgr0-dgr1-ii+1),tq(1:t+1-dgr0+dgr1),
                zeros(1,ii-1)];%shift qq1
            rr0=bitxor(rr0,tr);%add to rr0
            qq0=bitxor(qq0,tq);%add to qq0
        end%remainder has lower degree
        dgr0=find(rr0>0,1,'last');%find leading term
        fac=qin(rr0(dgr0));%inverse of leading coefficient
        rr0=mq(fac+1,rr0+1);%scale to make leading coefficient 1
        qq0=mq(fac+1,qq0+1);%scale qq0
        %rr0%result after each step
        if dgr0<t+1,
            flag=0;
        end %End Euclid
    end
    phase=bitxor(phase,1);%invert phase for next iteration
end

%%% Save the error locator, its degree, and the error evaluator

if zer==0, %nonzero syndrome
    if phase==1,
        qv=qq0;%error locator
        te=find(qq0>0,1,'last')-1;%expected number of errors
        rv=rr0(1:t+1); %error evaluator
        if dgr0>te,
            mer=1;%too many errors
        end
    else
        qv=qq1;%error locator
        te=find(qq1>0,1,'last')-1;%expeccted number of errors
        rv=rr1(1:t);%error evaluator
        if dgr1>te,
            mer=1;%too many errors
        end
    end
end
```

```
else
te=0;%no errors to be corrected
rv=zeros(1,t); qv=zeros(1,t+1);
end %Save the error locator and evaluator

%%%%%%%%%%%%%%%%%%%%%%%%%%%%%%%%%%%%%%%%%%%
%Search for errors
%Input: flags for zero errors, zer,
%        and too many errors, mer, Error locator, qv,
%        error evaluator, rv, expected number of errors, te
%Output: number of errors found, tf, error positions, pos,
%        error values, val
%%%%%%%%%%%%%%%%%%%%%%%%%%%%%%%%%%%%%%%%%%%

tf=0;%errors found
%Start search for error pos if zer and mer are 0
if zer+mer==0,
    ww=qex(1:t+1);
    % powers of alpha in the positions of the polynomials
    for ii=1:n,
        qv1=qv(1);
        for jj=3:2:t+1,
            qv1=bitxor(qv1,qv(jj));
        end% add even-order terms
        qv2=qv(2);
        for jj=4:2:t+1,
            qv2=bitxor(qv2,qv(jj));
        end% add odd-order terms, xq' value
        qv3=bitxor(qv1,qv2);
        if qv3==0,
            tf=tf+1;%zero found
            pos(tf)=ii;%store pos
            rv1=rv(1);
            for jj=2:t,
                rv1=bitxor(rv1,rv(jj));%rv value
            end
            val(tf)=mq(rv1+1,qin(qv2)+1);%store value
        end
        qn=mq(qv+1,ww+1);
        qv=diag(qn)';
        rn=mq(rv+1,ww+1);
        rv=diag(rn)';%next position
    end
end% Search for error positions and values
```

```
%%%%%%%%%%%%%%%%%%%%%%%%%%%%%%%%%%%%%%%%%%%%%%%
%Correct errors
%Input: Expected number of errors, te,
%number of errors found, tf, flag for zero errors, zer,
%flag for too many errors, mer, received word, rr,
%error positions, pos, error values, val
%Output: Corrected word, cout,
%flag for too many errors, mer, no. of corrected errors, tf
%%%%%%%%%%%%%%%%%%%%%%%%%%%%%%%%%%%%%%%%%%%%%%%

cout=rr;%output word
if zer+mer==0,
    if(te==tf),%correct number of errors has been found
    fac=qex(mod(-2*t*(pos-1),q-1)+1);
    for ii=1:te,
        val(ii)=mq(val(ii)+1,fac(ii)+1);%finish error value
        cout(n-pos(ii)+1)=bitxor(cout(n-pos(ii)+1),val(ii));
        %correct the output symbol
    end
    else
        mer=1;%errors not corrected
    end
end
cout %decoded word
mer
te
```

Index

Printed in the United States
by Baker & Taylor Publisher Services